Recent Developments in Lipid and Lipoprotein Research

A. Steinmetz J. Schneider
H. Kaffarnik (Eds.)

Hormones in Lipoprotein Metabolism

With 75 Figures and 50 Tables

Springer-Verlag
Berlin Heidelberg New York
London Paris Tokyo
Hong Kong Barcelona
Budapest

Priv.-Doz. Dr. Armin Steinmetz
Prof. Dr. Jürgen Schneider
Prof. Dr. Hans Kaffarnik

Klinikum der Philipps-Universität Marburg,
Zentrum für Innere Medizin,
Abt. für Endokrinologie und Stoffwechsel,
Baldingerstraße, D-W-3550 Marburg, FRG

ISBN-13:978-3-540-55995-5 e-ISBN-13:978-3-642-84855-1
DOI: 10.1007/978-3-642-84855-1

Library of Congress Cataloging-in-Publication Data
Hormones in lipoprotein meatabolism / A. Steinmetz, J. Schneider, H. Kaffarnik, editors, p. cm. Includes bibliographical references and index.
ISBN-13:978-3-540-55995-5
1. Lipoproteins-Metabolism-Congresses. 2. Hormones-Physiological effect-Congresses. I. Steinmetz, A. (Armin), 1952–. II. Schneider, J. (Jürgen). III. Kaffarnik, H. (Hans), 1929–.
QP552.L5H67 1993 612.3'98-dc20 92-34028 CIP

23/3145-5 4 3 2 1 0 – Gedruckt auf säurefreiem Papier

Preface

An intimate relation between hormones and lipoprotein metabolism has been known for a long time especially from hormone-deficiency or -overproduction syndromes. The mechanisms through which hormones influence lipid metabolism have become a field of major interest in science. The more we learn about cellular hormonal actions the better we understand regulatory processes and phenomena occuring in patients. The present book summarizes data discussed at an International Meeting at Marburg, Germany, on "Hormones in Lipoprotein Metabolism". It was an attempt to pull together knowledge in basic science from the mode of action of hormones all the way to clinical appearance of hormonal disorders involving lipoprotein metabolism. Thus data on molecular biology, on hormonal regulation of apolipoprotein synthesis, on lipoprotein receptors and enzyme induction are discussed together with the large field of oral contraceptive use, postmenopausal estrogen substitution, lipid disorders in diabetes mellitus, in thyroid dysfunction, in adrenal insufficiency and in glucocorticoid application. We are aware of the fact that such an overview cannot be complete but should serve as a collection of data and ideas for those interested in hormonal regulation of lipid metabolism. We thank all authors for taking an extra effort in writing up their presentations and thus making this edition possible. We also thank Bristol-Myers-Squibb for generous support of the meeting and of the final publication.

Marburg, Germany

ARMIN STEINMETZ
JÜRGEN SCHNEIDER
HANS KAFFARNIK

Contents

List of Authors

AMOUYEL, P.
Service de Sante Publique – Institut Pasteur de Lille, 1, Rue du Professeur Calmette, B.P. 245. F-59019 Lille Cedex, France

ASSMANN, G.
Institut für Klinische Chemie und Laboratoriumsmedizin, Universität Münster, D-W-4400 Münster, FRG

AUSIEKER, S.
Universität Mainz, III. Medizinische Klinik und Poliklinik, Endokrinologie, D-W-6500 Mainz, FRG

AUWERX, J.
LEGENDO, Katholieke Universiteit Leuven, Belgium

BÄNSCH, D.
Institut für Klinische Chemie und Laboratoriumsmedizin, Universität Münster, D-W-4400 Münster, FRG

BARD, J. M.
SERLIA & INSERM U325, Institut Pasteur 1, Rue du Professeur Calmette, B.P. 245, F-59019 Lille Cedex, France

BAUER, H.
Gynäkologisch-geburtshilfliche Praxis, D-W-3550 Marburg, FRG

BAUER, K.
Philipps-Universität, Zentrum Innere Medizin, Abt. Endokrinologie und Stoffwechsel, D-W-3550 Marburg, FRG

BAUMSTARK, M. W.
Medizinische Universitätsklinik, Abteilung Sport- und Leistungsmedizin, Hugstetter Str. 55, D-W-7800 Freiburg, FRG

BEATO, M.
Institut für Molekularbiologie und Tumorforschung, Emil-Mannkopf-Str. 2, D-W-3550 Marburg, FRG

BECKER, D.
Institut für Präventive Kardiologie der Medizinischen Universitätsklinik, D-W-6650 Homburg, FRG

BERG, A.
Medizinische Universitätsklinik, Abteilung Sport- und Leistungsmedizin, Hugstetter Str. 55, D-W-7800 Freiburg, FRG

BEYER, J.
Universität Mainz, III. Medizinische Klinik und Poliklinik, Endokrinologie, D-W-6500 Mainz, FRG
BIERMANN, J.
Sanatorium Ulrichsheim, Türkheimer Str. 4, D-W-8939 Bad Wörishofen
BORRMANN, M.
Institut für Klinische Chemie und Laboratoriumsdiagnostik, Medizinische Akademie „Carl Gustav Carus“, D-O-8090 Dresden, FRG
BRAITINGER, S.
Institut für Kernspintomographie, D-W-8390 Passau, FRG
BREWER, JR., H. B.
Molecular Disease Branch, National Heart, Lung, and Blood Institute, National Institutes of Health, 9000 Rockville Pike, Bethesda, Maryland, USA
BUNTE, T.
Institut für Präventive Kardiologie der Medizinischen Universitätsklinik, D-W-6650 Homburg, FRG
DABELS, J.
Medizinische Fakultät, Innere Medizin, Endokrinologie und Stoffwechsel, Universität Rostock, D-O-2500 Rostock, FRG
DANULL, E.
Department of Medicine II, University of Cologne, Ostmerheimer Str. 200, D-W-5000 Köln 91, FRG
DAUME, E.
Medizinisches Zentrum für Frauenheilkunde und Geburtshilfe, Klinik für gynäkologische Endokrinologie und Reproduktion, Universität Marburg, Pilgrimstein 3, D-W-3550 Marburg
DIRKES-KERSTING, A.
Institut für Klinische Chemie und Laboratoriumsmedizin, Universität Münster, D-W-4400 Münster, FRG
ECKARDT, H.
Department of Medicine II,University of Cologne, Ostmerheimer Str. 200, D-W-5000 Köln 91, FRG
v. ECKARDSTEIN, A
Institut für Arterioskleroseforschung an der Universität Münster, Domagkstr. 3, D-W-4400 Münster, FRG
EGGERMAN, T.
Molecular Disease Branch, National Heart, Lung, and Blood Institute, National Institutes of Health, 9000 Rockville Pike, Bethesda, Maryland, USA
EHLENZ, K.
Philipps Universität, Zentrum Innere Medizin, Abt. Endokrinologie und Stoffwechsel, D-W-3550 Marburg, FRG

FAIRWELL, T.
Molecular Disease Branch, National Heart, Lung, and Blood Institute, National Institutes of Health, 9000 Rockville Pike, Bethesda, Maryland, USA

FINGER, E.
Medizinische Fakultät, Innere Medizin, Endokrinologie und Stoffwechsel, Universität Rostock, D-O-2500 Rostock, FRG

FRUCHART, J. C.
SERLIA & INSERM U325, Institut Pasteur, 1, Rue du Professeur Calmette, B.P. 245, F-59019 Lille Cedex, France

GREGG, R. E.
Squibb Institute for Medical Research, Princeton, NJ 08543, USA

GROSS, W.
Zentrum für Biologische Chemie, Klinikum der Johann-Wolfgang Goethe Universität, Theodor-Stern-Kai 7, D-W-6000 Frankfurt a.M. 70, FRG

HAHMANN. H.
Institut für Präventive Kardiologie der Medizinischen Universitätsklinik, D-W-6650 Homburg, FRG

HAZZARD, W. R.
Department of Internal Medicine, J. Paul Sticht-Center on Aging, Bowman Gray School of Medicine, Wake Forest University, Winston-Salem, NC 27103, USA

HEINRICH, J.
Institut für Klinische Chemie und Laboratoriumsmedizin, Universität Münster, D-W-4400 Münster, FRG

HERRMAN, W.
Institut für klinische Chemie, Universität Regensburg, Franz-Josef-Strauß-Allee 11, D-W-8400 Regensburg, FRG

HOEG, J. M.
Molecular Disease Branch, National Heart, Lung, and Blood Institute, National Institutes of Health, 9000 Rockville Pike, Bethesda, Maryland, USA

JÄCKLE, S.
Medizinische Kernklinik und Poliklinik, Universitätskrankenhaus Eppendorf, Martinistr. 52, D-W-2000 Hamburg 20, FRG

JACOB, R.
Department of Medicine II, University of Cologne, Ostmerheimer Str. 200, D-W-5000 Köln 91, FRG

JAKOB, E.
Medizinische Universitätsklinik, Abteilung Sport- und Leistungsmedizin, Hugstetter Str. 55, D-W-7800 Freiburg, FRG

JAROSS, W.
Medizinische Akademie „Carl Gustav Carus", Institut für klinische Chemie und Laboratoriumsdiagnostik, Fetscherstr. 74, D-O-8090 Dresden, FRG

JUNG-HOFFMANN, C.
Zentrum für Gynäkologie und Geburtshilfe, Klinikum der Johann-Wolfgang Goethe-Universität, Theodor-Stern-Kai 7, D-W-6000 Frankfurt 70, FRG

KAFFARNIK, H.
Philipps Universität, Zentrum Innere Medizin, Abt. Endokrinologie und Stoffwechsel, D-W-3550 Marburg, FRG

KELLER, H.-E.
Klinisch-Chemisches Zentrallaboratorium der Medizinischen Universitätsklinik, D-W-6650 Homburg, FRG

KEUL, J.
Medizinische Universitätsklinik, Abteilung Sport- und Leistungsmedizin, Hugstetter Str. 55, D-W-7800 Freiburg, FRG

KINDT, M. R.
Molecular Disease Branch, National Heart, Lung, and Blood Institute, National Institutes of Health, 9000 Rockville Pike, Bethesda, Maryland, USA

KLOTZ, B.
Institut für Präventive Kardiologie der Medizinischen Universitätsklinik, D-W-6650 Homburg, FRG

KOHRING, S.
Institut für Präventive Kardiologie der Medizinischen Universitätsklinik, D-W-6650 Homburg, FRG

KRAUSE, U.
Universität Mainz, III. Medizinische Klinik und Poliklinik, Endokrinologie, D-W-6500 Mainz, FRG

KUHL, H.
Zentrum für Gynäkologie und Geburtshilfe, Klinikum der Johann-Wolfgang Goethe-Universität, Theodor-Stern-Kai 7, D-W-6000 Frankfurt 70, FRG

LACKNER, K. J.
Institut für klinische Chemie, Universität Regensburg, Franz-Josef-Strauß-Allee 11, D-W-8400 Regensburg, FRG

LAUE, C.
Universität Mainz, III. Medizinische Klinik und Poliklinik, D-W-6500 Mainz, FRG

LESCHKE, M.
Universität Düsseldorf, Zentrum Innere Medizin, Kardiologie, D-W-4000 Düsseldorf, FRG

LINDHOFER, H.-G.
Krankenhaus, Ernststr. 7-9, D-O-6100 Meiningen, FRG

LULEY, C.
Medizinische Universitätsklinik, Zentrallabor, Hugstetter Str. 55, D-W-7800 Freiburg, FRG

MAISCH, B.
Zentrum Innere Medizin, Schwerpunkt Kardiologie, Philipps-Universität Marburg, D-W-3550 Marburg, FRG
MALLE, E.
Philipps Universität, Zentrum Innere Medizin, Abt. Endokrinologie und Stoffwechsel, D-W-3550 Marburg, FRG
MÄRZ, W.
Zentrum für Biologische Chemie, Klinikum der Johann-Wolfgang Goethe Universität, Theodor-Stern-Kai 7, D-W-6000 Frankfurt a.M. 70, FRG
PATTERSON, A.
Molecular Disease Branch, National Heart, Lung, and Blood Institute, National Institutes of Health, 9000 Rockville Pike, Bethesda, Maryland, USA
PLACH, E.
I. Medizinische Klinik am Zentrum für Innere Medizin des Klinikums Passau, D-W-8390 Passau, FRG
PRELLWITZ, W.
Universitätsklinik Mainz, Zentrallabor, D-W-6500 Mainz, FRG
PROBST, R.
Universität Mainz, III. Medizinische Klinik und Poliklinik, Endokrinologie, D-W-6500 Mainz, FRG
RADER, D. J.
Molecular Disease Branch, National Heart, Lung, and Blood Institute, National Institutes of Health, 9000 Rockville Pike, Bethesda, Maryland, USA
RATZMANN, K.-P.
Zentralstelle für Diabetes und Stoffwechselkrankheiten, D-O-1000 Berlin, FRG
RICHTER, W. O.
Medical Department II, Klinikum Großhadern, Ludwig-Maximilians-University, Marchioninistr. 15, D-W-8000 München 70, FRG
RITTER, M. M.
Medical Department II, Klinikum Großhadern, Ludwig-Maximilians-University, Marchioninistr. 15, D-W-8000 München 70, FRG
SANDKAMP, M.
Institut für Klinische Chemie und Laboratoriumsmedizin, Universität Münster, D-W-4400 Münster, FRG
SARTOR, H.
Medizinische Klinik der Universität, Abteilung für Innere Medizin III, Bergheimer Str. 58, D-W-6900 Heidelberg, FRG
SCHAEFER, J. R.
Philipps Universität, Zentrum Innere Medizin, Abteilung Endokrinologie und Stoffwechsel, D-W-3550 Marburg

SCHÄTZER-KLOTZ, D.
Institut für Präventive Kardiologie der Medizinischen Universitätsklinik, D-W-6650 Homburg, FRG
SCHIEFFER, H.
Institut für Präventive Kardiologie der Medizinischen Universitätsklinik, D-W-6650 Homburg, FRG
SCHNEIDER, J.
Philipps-Universität, Zentrum Innere Medizin, Abt. Endokrinologie und Stoffwechsel, D-W-3550 Marburg
SCHREZENMEIR, J.
Universität Mainz, III. Medizinische Klinik und Poliklinik, Endokrinologie, D-W-6500 Mainz, FRG
SCHULTE, H.
Institut für Klinische Chemie und Laboratoriumsmedizin, Universität Münster, D-W-4400 Münster, FRG
SCHWANDT, P.
Medical Department II, Klinikum Großhadern, Ludwig-Maximilians-University, Marchioninistr. 15, D-W-8000 München 70, FRG
SIEKMEIER, R.
Zentrum für Biologische Chemie, Klinikum der Johann-Wolfgang Goethe Universität, Theodor-Stern-Kai 7, D-W-6000 Frankfurt a.M. 70, FRG
SLATER, E. P.
Institut für Molekularbiologie und Tumorforschung, Emil-Mannkopf-Str. 2, D-W-3550 Marburg, FRG
SOENNICHSEN, A. C.
Medical Department II, Klinikum Großhadern, Ludwig-Maximilians-University, Marchioninistr. 15, D-W-8000 München 70, FRG
STAELS, B.
LEGENDO, Katholieke Universiteit Leuven, Belgium
v. STEIN, T.
Medizinische Universitätsklinik, Abteilung Sport- und Leistungsmedizin, Hugstetter Str. 55, D-W-7800 Freiburg, FRG
STEINMETZ, A.
Philipps Universität, Zentrum Innere Medizin, Abt. Endokrinologie und Stoffwechsel und Medizinische Poliklinik, D-W-3550 Marburg, FRG
STEINSEIFER, J.
Philipps Universität, Zentrum Innere Medizin, Abt. Endokrinologie und Stoffwechsel, D-W-3550 Marburg, FRG
STELLWAAG, M.
Zentrum Innere Medizin, Schwerpunkt Kardiologie, Philipps-Universität Marburg, D-W-3550 Marburg, FRG
STRAUER, B. E.
Zentrum Innere Medizin, Kardiologie, Universität Düsseldorf, D-W-4000 Düsseldorf, FRG

TATÒ, F.
Universität Mainz, III. Medizinische Klinik und Poliklinik, Endokrinologie, D-W-6500 Mainz, FRG
TATÒ, S.
Universität Mainz, III. Medizinische Klinik und Poliklinik, Endokrinologie, D-W-6500 Mainz, FRG
TEICHMANN, A. T.
Klinikum Aschaffenburg, Frauenklinik, Am Hasenkopf, D-W-8750 Aschaffenburg, FRG
TRÜBSBACH, A.
Institut für Klinische Chemie und Laboratoriumsdiagnostik, Medizinische Akademie „Carl Gustav Carus“, D-O-8090 Dresden, FRG
WAMBACH, G.
St. Elisabeth Hospital, Am Schloßpark 12, D-W-4353 Herten 1, FRG
WEBER, L.
I. Medizinische Klinik am Zentrum für Innere Medizin des Klinikums Passau, D-W-8390 Passau, FRG
WEBER, P.
Universität Mainz, III. Medizinische Klinik und Poliklinik, Endokrinologie, D-W-6500 Mainz, FRG
WINDLER, E.
Medizinische Kernklinik und Poliklinik, Universitäts-Krankenhaus Eppendorf, Martinistr. 52, D-W-2000 Hamburg 20, FRG
WOLTER, G.
Medizinische Klinik der Universität, Abteilung für Innere Medizin III, Bergheimer Str. 58, D-W-6900 Heidelberg, FRG
ZECH, L. A.
Molecular Disease Branch, National Heart, Lung, and Blood Institute, National Institutes of Health, 9000 Rockville Pike, Bethesda, Maryland,
ZEHNER, J.
I. Medizinische Klinik am Zentrum für Innere Medizin des Klinikums Passau, D-W-8390 Passau, FRG

Interaction of Steroid Hormone Receptors with DNA

E. P. SLATER and M. BEATO

Steroid hormones bind to their corresponding receptors in the cytoplasm of responsive cells and this complex is translocated to the nucleus where the hormone-receptor-complex binds to specific DNA sequences located in the regulatory regions of the target gene. These specific DNA sequences are referred to as hormone responsive elements (HRE) and act as enhancer elements such that when the receptor complex binds, transcription is induced. In the absence of ligand the hormone receptor is maintained in an inactive form by association with other cellular proteins.

As a model system for induction of transcription by steroid hormone receptors, we have chosen the mouse mammary tumor virus (MMTV). Induction requires binding of the hormone receptor to an HRE that is organized in chromatin. Following receptor binding, changes in chromatin structure occur that then allow binding of transcription factors to the MMTV promoter or regulatory region of the gene. One of these factors is nuclear factor I which is important for basal transcription. The factor that is thought to mediate the induction by steroid hormones is the octamer transcription factor, OTF-1. Mutation of the octamer binding sites completely abolishes the stimulatory effect of the progesterone receptor.

Glucocorticoids also inhibit expression of many genes. Two mechanisms for this phenomenon will be discussed. The effect on the gene for the α-subunit of chorionic gonadotropin is partially due to DNA binding competition between the receptor and another transcription factor (CREB) for the gene. A second mechanism observed in the repression of the collagenase gene involves an interaction of the receptor with components of the AP1 complex, Jun and Fos.

These various examples of regulation by steroid hormones will be discussed.

Structure of the Hormone Receptor

In the past few years most of the genes for the steroid hormone receptors have been cloned and sequenced. This led to the finding that steroid hormone receptors belong to a large superfamily of nuclear receptors which includes the receptors for retinoic acid, thyroid hormones, and several genes for which a physiological ligand is not yet known.

The genes for all members of the nuclear receptor family have common features summarized in Fig. 1 [2]. Beginning at the amino terminus, there is the A/B or Modulator domain, the nuclear receptor's most variable region. The function of this domain is also less well-defined. A transactivation function, which seems to be independent of ligand binding, has been assigned to this region. Whether this region plays a role in the synergistic interactions among receptor dimers bound to DNA is not yet clear. In the case of the progesterone receptor in humans, rodents and chicken, several forms of the receptor are found that differ in the length of this amino terminal region. There are indications that these alternate forms may differ in their ability to induce different genes, and that these effects could also be cell-type-specific.

In the middle of the protein is the C or DNA binding domain. This region contains a short DNA-binding domain composed of 70 amino acid residues including many conserved cysteines. Eight of these cysteines are thought to be organized into two "zinc fingers," each containing four cysteine residues tetrahedrally coordinating a zinc ion. A comparison of the amino acid sequences in the DNA-binding domain of the different nuclear receptor genes resulted in the classification of the various receptors into two subfamilies [2]. The glucocorticoid receptor (GR) is the prototype of the smaller subfamily that includes the progesterone receptor (PR), the androgen receptor and the mineralocorticoid receptor. The prototype of the larger subfamily is the estrogen receptor (ER) and this group includes the vitamin D3 receptor, the various thyroid hormone receptors, the receptor for retinoic acid and many of the "orphan" receptors. The main differences between the members of the two subgroups of nuclear receptors reside in the knuckles of the two zinc fingers, in regions that have been shown to be important for DNA sequence recognition and receptor dimerization. The amino acids following the first finger have been shown to be organized into an α-helix that penetrates the major

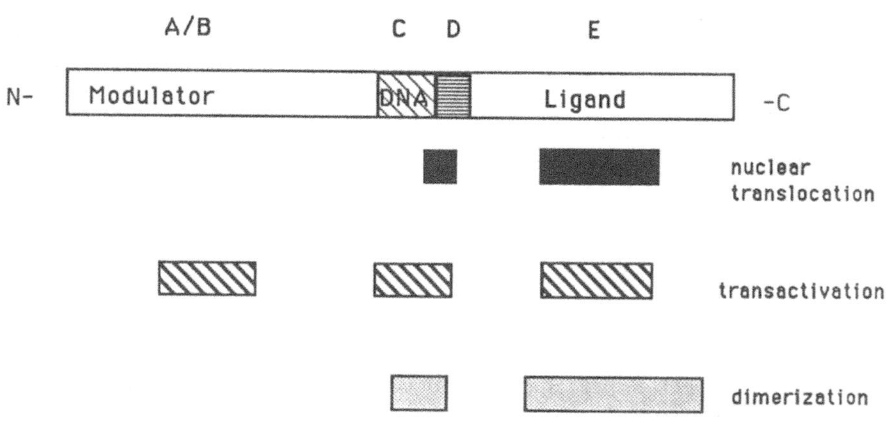

Fig. 1. Structure of Nuclear Receptors

groove and recognizes the specific nucleotide sequences of the hormone responsive element (HRE). Following the second zinc finger there is another α-helical region that serves to position the recognition helix in the correct orientation. Therefore, the general structure of the DNA binding domain of the nuclear receptors is reminiscent of the structure found in many other prokaryotic and eukaryotic DNA binding proteins, namely: helix, loop/turn, helix.

The D and E domains contain nuclear translocation functions. The large carboxy-terminal domain is responsible for binding of the hormone ligand. In the GR and PR some of the amino acid residues that are contacted by the hormone ligand have been identified in photo-crosslinking experiments. Genetic analysis of the ER has also been used to map the region of the hormone-binding domain responsible for direct binding of estrogens. In these latter studies, it was possible to distinguish the region implicated in hormone binding from that responsible for ligand-induced receptor dimerization. The hormone binding domain also contains a transcriptional activation function that is dependent for activity on binding of the ligand [2].

Hormone Responsive Elements

The nucleotide sequences recognized by the steroid hormone receptors are responsible for their effects on the gene activity, the hormone-responsive or regulatory elements (HRE) have been studied in great detail. In most cases, these sequences have a palindromic structure with two unequal halves separated by three non-conserved base pairs. Just as the nuclear receptors can be divided into two subgroups according to the structure of their DNA-binding domains mentioned above, the HREs can also be divided into two subgroups: the GRE/PRE subgroup and the ERE subgroup.

The half palindrome of the GRE/PRE has the general structure TGTYCT, whereas the prototype ERE half is TGACC (Fig. 2). We have recently shown that the last base of the half GRE/PRE, though highly conserved, is not essential for receptor binding nor for functional activity.[13] Therefore, the main differences between the two types of HREs reside in the third and fourth positions. As the fourth position of the GRE/PRE is a C in 49% of the cases, the most important distinction is the T in the third position of GRE/PRE as opposed to the A of the ERE. We know that GR and PR contact the 5′-methyl group of the T in the third position, but this base can be replaced by an A, provided that a T is located in the fourth position. More-

GRE / PRE GGTACAnnnTGTYCT

1 2 3 4 5 6

ERE AGGTCAnnnTGACCT

Fig. 2. Consensus Sequences for Responsive Elements (Y = C or T)

over, the ERE of the rabbit uteroglobin promoter contains a C in the third position [12], suggesting the ER does not directly contact this position of the ERE. It seems that the sequence context of the individual positions allows some flexibility in the way that the DNA binding domain of the receptor contacts the HRE. This flexibility is further documented by the observation that hybrid HREs composed of a half-GRE/PRE linked to a half-ERE are recognized by GR and PR as well as by ER, and respond to glucocorticoids, progestins and estrogens in transfection experiments [14].

The amino acid side chains responsible for the distinction between GRE/PRE and ERE are apparently located at the knuckle of the first zinc finger, between cysteines 3 and 4 and immediately downstream in the recognition helix. In addition, a few amino acid residues located between the cysteines 1 and 2 of the second finger are also important in specifying HRE target recognition, probably by dictating the precise orientation of receptor dimers. As the three dimensional structure of the zinc finger region of GR has been elucidated in NMR-studies [7], it will soon be possible to postulate and to test direct contacts between individual amino acid side chains and individual base pairs within the HRE.

Transcriptional Activation

MMTV-HRE

The first identification of DNA elements able to mediate hormone induction in gene transfer experiments was accomplished with the MMTV system (for review [2]). In addition, the first demonstration of specific binding of steroid hormone receptors to DNA was performed with this system. In the GR mouse strain there are four copies of the hexanucleotide motif TGTTCT (GRE/PRE) between position –190 and –75 of the MMTV promoter and each of these motifs is contacted by GR or PR [5].

Binding of several receptor molecules, and their functional interaction are known to be essential for induction of MMTV transcription. To determine the mechanism of the transactivation, the factors that mediate the effect of receptor binding on transcriptional efficiency had to be identified. The participation of the transcription factor nuclear factor 1 (NF1) in hormonal induction of the MMTV promoter was demonstrated in complementation experiments with an NF1-deficient choriocarcinoma cell line. However, binding of NF1 to its cognate sequence between –75 and –63 on the MMTV promoter is not favored by binding of the hormone receptors. On the contrary, both proteins compete for DNA binding *in vitro*, in agreement with the overlap of their respective binding sites [4]. Therefore, a simple mechanism involving binding cooperativity between hormone receptors and NF1 is not in agreement with these findings.

In cell-free transcription experiments, Kalff et al. [8] demonstrated that addition of the PR to templates containing the MMTV promoter enhances their transcriptional efficiency about 10-fold. In this system, deletion of the NF1 binding site, or addition of an excess of oligonucleotide carrying the NF1 consensus sequence, reduced the basal expression of the MMTV promoter dramatically, but did not influence the stimulatory effect of preincubation with the PR. These *in vitro* data confirmed the *in vivo* results and in addition show that NF1 acts as a basal transcription factor on the MMTV promoter, without any indication for synergism or cooperativity with the hormone receptors. The conclusion is that factors other than NF1 must mediate the transactivation of the MMTV promoter by hormone receptors.

In contrast, mutation of the octamer motifs located between the NF1 binding sites and the TATA box resulted in a significant reduction in the hormonal induction of the MMTV promoter in gene transfer experiments [3]. These mutated promoters were completely unresponsive to the addition of PR *in vitro*, suggesting that the effect of PR on cell-free transcription of the MMTV promoter is actually mediated by the octamer binding factor. Since OTF-1 is the main octamer binding factor in HeLa cells, from which the nuclear extracts are prepared, and purified OTF-1 binds to the two octamer motifs of the MMTV promoter, the assumption is that the effect of PR on transcription of MMTV-DNA is mediated by OTF-1 [3].

The lack of cooperativity between hormone receptors and NF1 in terms of binding to DNA was unexpected, because *in vivo* hormone treatment induces NFl binding to the MMTV promoter [6]. On the other hand, the MMTV-LTR is known to be precisely organized into nucleosomes and following hormone administration a DNaseI hypersensitive site appears over the HRE [15]. Thus it seemed plausible that changes in chromatin structure could mediate the effect of hormone treatment on NF1 binding. The position of the DNA double helix on the surface of the histone octamer could account for the lack of binding of NF1 prior to hormone treatment. In the absence of hormone, the promoter is inaccessible to NF1 and silent in vivo. If the model is correct, and promoter occlusion is due to nucleosome positioning, binding of receptor to the nucleosomally organized MMTV promoter should alter its structure and expose the recognition sequence of NF1. Upon binding of the receptor, the nucleosome is not disassembled, but the region containing the NF1 binding site becomes more accessible to digestion by exonuclease [11]). This model would also explain why transcription of the naked MMTV promoter *in vitro* is rather efficient even in the absence of the hormone receptor [8].

Hormonal Repression

Steroid hormones are also able to repress the expression of certain genes. For example, glucocorticoids are clinically used due to their suppressive effect on inflammation and immune response. Once the basic mechanism of gene activation by steroid hormones became clear, the question concerning the mechanism of gene repression was formulated more precisely. Cloning of the receptors demonstrated that for most of the steroid hormones there is only one class of receptor molecules, that should mediate all positive as well as negative effects of the corresponding hormones. Therefore, the question arose as to whether the same type of DNA regulatory elements that mediate the gene induction is also responsible for repression. Two possible mechanisms for hormonal repression will be presented here (Fig. 3).

Among the first glucocorticoid repressed genes to be studied was the gene for the α-subunit of the glycoprotein hormones. The expression of this gene in pituitary and placental cells is stimulated by cAMP and is inhibited by glucocorticoids at the transcriptional level [1]. Induction by cAMP is mediated by two copies of the cAMP regulatory element (CRE) that are recognized by a specific protein, the CRE-binding protein (CREB). In transfection experiments, the negative effect of glucocorticoids is only observed with cells in which the CRE is active, whereas in cells not responding to cAMP, glucocorticoids have a stimulatory effect on the α-subunit promoter. These findings suggest that the effect of glucocorticoids is to counteract the induction by cAMP. In agreement with this concept, DNA binding experiments with purified GR identified three binding sites on the α-subunit promoter, two of which overlap with the CREs [1]. Moreover, DNA binding is the only required function of the GR; the amino-terminal and carboxy-terminal halves can be removed or replaced by β-galactosidase without impairing repression

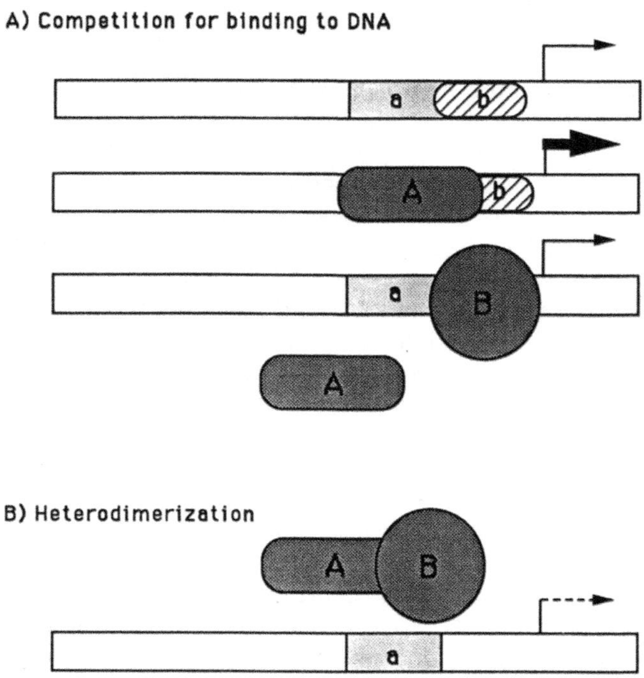

Fig. 3. Mechanisms of Transcriptional Repression

[10]. Thus, in this particular case, glucocorticoid repression could be due to competition by GR and positive factors for overlapping target sites on the promoter (Fig. 3A).

An apparently different mechanism of repression is operative in the glucocorticoid inhibition of AP-1 mediated induction [9].

Binding of GR to DNA is not essential for transrepression. Rather, GR seems to inhibit the activation of the Fos/Jun complex by some kind of interaction with the AP-1 complex. This idea is supported by the reverse effect, namely the inhibition of GR-mediated induction of a GRE by Fos, in the absence of Fos binding to the GRE [9]. One possible mechanism to explain this type of repression would be to postulate a heterodimerization between a receptor monomer and other components of the AP-1 complex (Fig. 3B). This type of cross-talk between nuclear receptors and nuclear factors involved in other signal transduction pathways probably exemplifies a common behavior of transregulatory proteins.

Conclusions

This review of the MMTV promoter suggests that there are at least two different mechanisms by which promoter activation can take place. One involves removal of promoter repression due to nucleosome positioning, and is mediated mainly by NF1. In this mechanism there is no direct interaction between the hormone receptors and the responsive transcription factor. The second mechanism follows a more conventional pathway. The receptors bound to the HRE facilitate the interaction of OTF-1 with the two octamer motifs of the MMTV promoter.

Two mechanisms of repression were also discussed: one involving DNA binding resulting in competition with transcription factors for their binding sites on the DNA and the other possibly resulting from a protein-protein interaction of two different transactivators resulting in a heterodimer that would be inactive.

Acknowledgements. The experimental work summarized in this article was supported by grants form the Deutsche Forschungsgemeinschaft and the Fonds der Chemischen Industrie.

References

1. Akerblom IE, Slater EP, Beato M, Baxter JD, Mellon PL (1988) Negative Regulation by Glucocorticoids Through Interference with a cAMP Responsive Enhancer. Science 241:350–353.
2. Beato M (1989) Gene Regulation by Steroid Hormones. Cell 56:335–344.
3. Bruggemeier U, Kalff M, Franke S, Scheidereit C, Beato M (1991) Ubiquitous Transcription Factor OTF-l Mediates Induction of the MMTV Promoter through Synergistic Interaction with Hormone Receptors. Cell 64:565–572.

4. Bruggemeier U, Rogge L, Winnacker EL, Beato M (1990) Nuclear factor I acts as a transcription factor on the MMTV promoter but competes with steroid hormone receptors for DNA binding. EMBO J 9:2233–2239.
5. Chalepakis G, Arnemann J, Slater EP, Bruller H-J, Gross B, Beato M (1988)Differential gene activation by glucocorticoids and progestins through the hormone regulatory element of mouse mammary tumor virus. Cell 53:371–382.
6. Cordingley MG, Riegel AT, Hager GL (1987) Steroid-dependent interaction of transcription factors with the inducible promoter of mouse mammary tumor virus in vivo. Cell 48:261–270.
7. Hard T, Kellenbach E, Boelens R, Maler BA, Dahlman K, Freedman LP, Carlsted-Duke J, Yamamoto KR, Gustafsson J, Kaptein R (1990) Solution structure of the glucocorticoid receptor DNA-binding domain. Science 199:157–160.
8. Kalff M, Gross B, Beato M (1990) Progesterone receptor stimulates transcription of mouse mammary tumor virus in a cell-free system. Nature 344:360–362.
9. Lucibello FC, Slater EP, Jooss KU, Beato M, Muller R (1990) Mutual transrepression of Fos and the glucocorticoid receptor: involvement of a functional domain in Fos which is absent in FosB. EMBO J 9:2827–2834.
10. Oro AE, Hollenberg SM, Evans RM (1988) Transcriptional inhibition by a glucocorticoid receptor-J3-galactosidase fusion protein. Cell 55:1109–1114.
11. Piña B, Bruggemeier U, Beato M (1990) Nucleosome positioning modulates accessibility of regulatory proteins of the mouse mammary tumor virus promoter. Cell 60:719–725.
12. Slater EP, Redeuilh G, Theis K, Suske G, Beato M (1990) The uteroglobin promoter contains a non-canonical estrogen responsive element. Mol Endocrinol 4:604–610.
13. Truss M, Chalepakis G, Beato M (1990) Contacts between steroid hormone receptors and thymines in DNA: An interference method. Proc Natl Acad Sci USA 87:7180–7184.
14. Truss M, Chalepakis G, Slater EP, Mader S, Beato M (1991) Functional Interaction of Hybrid Response Elements with Wild-Type and Mutant Steroid Hormone Receptors. Mol Cell Biol 11:3247–3258.
15. Zaret KS, Yamamoto KR (1984) Reversible and Persistent Changes in Chromatin Structure Accompany Activation of a Glucocorticoid-dependent Enhancer Element. Cell 38:29–38.

Hormonal Control of Genes Involved in Lipoprotein Metabolism in the Rat

B. STAELS and J. AUWERX

Hyperlipidemia is recognized as a major risk factor for atherosclerosis. In plasma lipids are carried by lipoproteins. The metabolism of these lipoproteins is dependent upon the interaction of their protein constituents, the apolipoproteins, with a variety of enzymes and receptors present in plasma or on the surface of cells in different organs and tissues. The level of expression of the genes coding for proteins involved in lipoprotein metabolism largely determines the metabolic fate of plasma lipoproteins. The major goal of these studies was to define the discrete regulation of these genes by a number of hormones known to affect lipoprotein metabolism such as estrogens, thyroid hormones and glucocorticoids. At the same time, we have explored whether changes in gene expression are paralleled by changes in the concentration or activity of the proteins involved.

Liver and intestinal RNA was extracted from rats treated with the indicated hormones (Chirgwin et al. 1979). The expression of the genes coding for apolipoproteins (apo A-I, apo A-II, apo A-IV, apo B and apo E), for the LDL-receptor, for hepatic lipase and for lipoprotein lipase was analyzed by measuring the corresponding mRNA levels using dot blot and northern blot hybridization procedures (Staels et al. 1989). Changes in gene expression were subsequently compared to the plasma concentrations or to the tissue activities of the corresponding proteins. Plasma or serum apolipoprotein concentrations were analyzed by electroimmunoassays (Dallinga-Thie et al. 1985). Plasma apo E increases nearly 2-fold, but liver apo E mRNA remains constant in hypothyroid rats (Staels et al. 1990c). In liver, hepatic lipase and LDL receptor mRNA levels decrease by 20% and nearly 50%, respectively (Staels et al. 1990c). Liver apo A-I and apo A-IV mRNA levels are lowered after treatment with n-propylthiouracil, but the decrease of apo A-IV mRNA is much more pronounced (Staels et al. 1990a & 1990c). In contrast, hepatic apo A-II mRNA levels increase 2-fold.

Hepatic apo A-I, apo A-II, apo A-IV, apo B and LDL-receptor mRNA levels change in the opposite direction in rats made hyperthyroid by L-thyroxine administration, whereas apo E and hepatic lipase mRNA levels remain unchanged (Figure 3; Staels et al. 1990c). None of the treatments causes major changes in intestinal apolipoprotein mRNA levels.

Both in humans and in laboratory animals, such as the rat, administration of glucocorticoids has been reported to affect plasma lipid concentrations. When male rats were treated with hydrocortisone a decrease in plasma cholesterol concentration is observed, whereas plasma triglycerides remain unchanged (Figure 4; Staels et al. 1991). Most intriguing, however, are the discordant effects observed between the natural hormone, hydrocortisone, and the synthetic hormones, dexamethasone and triamcinolone. Indeed, administration of triamcinolone and dexamethasone at comparable doses respectively increases plasma cholesterol and triglycerides (Staels et al. 1991). Small effects on apo E and apo B mRNA levels are observed, but all corticosteroids tested increase apo A-I and apo A-IV mRNA and decrease apo A-II mRNA levels in liver (Staels et al. 1991). Triamcinolone and dexamethasone, however, are 3 times more efficient in stimulating hepatic apo A-IV gene expression than hydrocortisone. In contrast, liver apo A-I and apo A-II mRNA levels are altered to a similar extent by the three corticosteroids (Staels et al. 1991). Plasma apo A-I and apo B concentrations reflect their respective liver mRNA levels after the distinct glucocorticoids used (Staels et al. 1991). For apo A-IV and apo E discrepancies between plasma and liver mRNA levels after different corticosteroids, however, point to additional regulatory effects

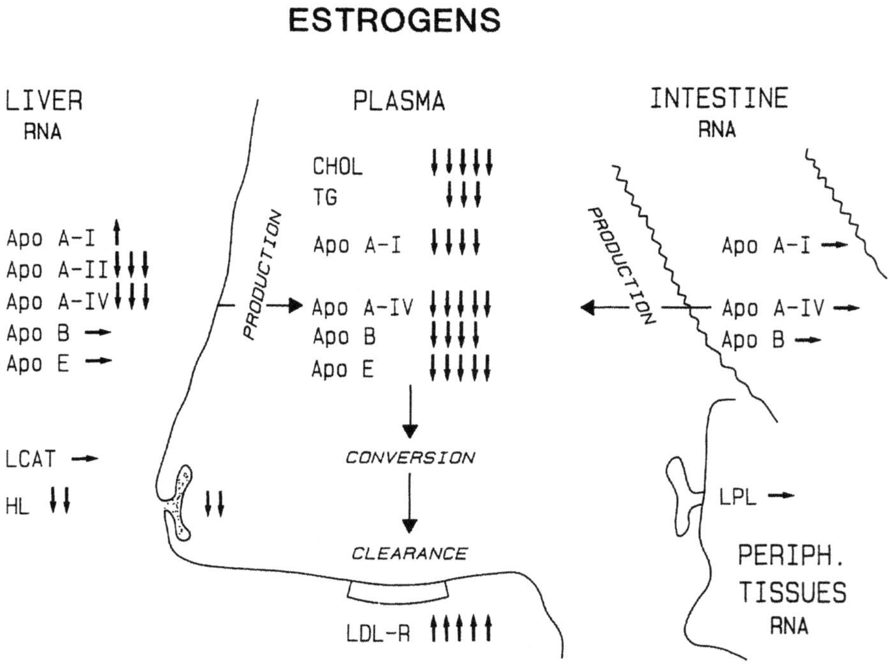

Fig. 1. Influence of estrogens (ethinylestradiol) on plasma lipid and apolipoprotein concentrations and mRNA levels for apolipoproteins (apo), lecithin:cholesterol acyltransferase (LCAT), hepatic lipase (HL), lipoprotein lipase (LPL) and the LDL-receptor (LDL-R) in rat liver, intestine and adipose tissue. Changes are semi-quantitatively expressed relative to the levels in sham-treated control rats (1 arrow = less than 2-fold difference from controls, 2 arrows = 2-fold difference, 3 arrows = 3-fold difference, etc...). For more details see Staels et al. 1989, 1990a & 1990b

on plasma apolipoprotein levels (Staels et al. 1991). Hepatic lipase mRNA and activity, remain constant after hydrocortisone (Figure 4), but decrease after triamcinolone and dexamethasone. Since this decrease is much more pronounced after triamcinolone, changes in hepatic lipase activity may contribute to the distinct plasma cholesterol concentrations after the different corticosteroids.

From these studies it can be concluded that the expression of the genes involved in plasma lipoprotein metabolism is regulated in a complex fashion. Different subsets of genes are affected by several hormones. Alterations in the synthesis of the apolipoproteins may result in changes in lipoprotein production, whereas changes in the expression of the LDL-receptor, hepatic lipase and lipoprotein lipase may influence the metabolism and the clearance of the lipoprotein particles from plasma. The final changes in plasma lipoprotein concentrations reflect the delicate balance of the interplay between synthesis and clearance. In a number of conditions the observed alterations in the concentrations of lipids and apolipoproteins parallel the observed changes in gene expression. For instance, the alterations in hepatic apo A-I gene expression after treatment with thyroid hormones (Figure 3) and corticoste-

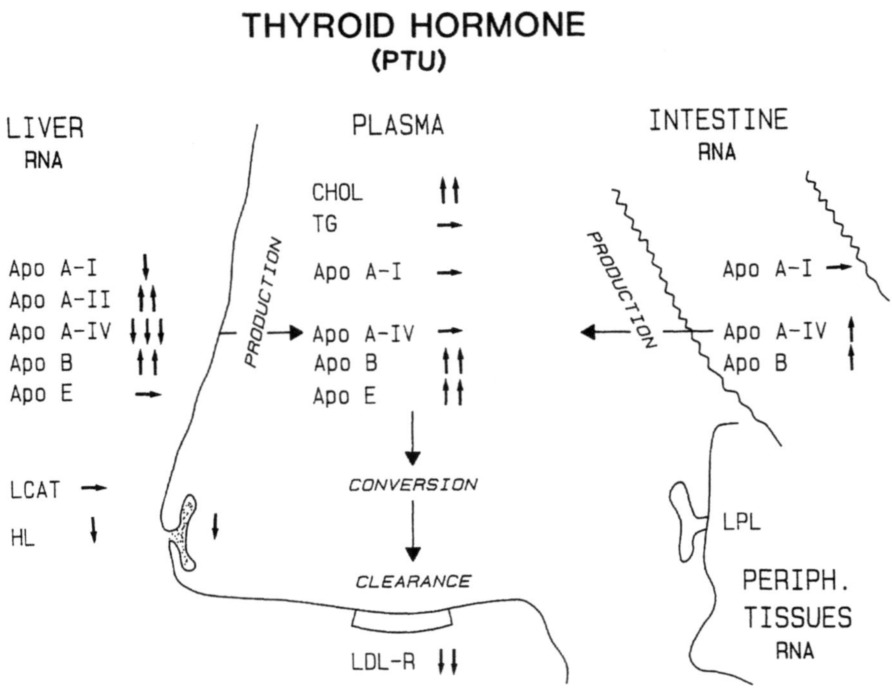

Fig. 2. Influence of n-propylthiouracil (PTU) -induced hypothyroidism on plasma lipid and apolipoprotein concentrations and mRNA levels for apolipoproteins (apo), lecithin:cholesterol acyltransferase (LCAT), hepatic lipase (HL), and the LDL receptor (LDL-R) in rat liver and intestine. Changes are expressed in a similar fashion as described in figure 1. For more details see Staels et al. 1990a & 1990c

roids (Figure 4) probably result in similar changes in apo A-I synthesis and secretion by the liver, leading to the observed alterations in plasma apo A-I concentrations. Similarly, plasma apo A-IV concentrations parallel changes in hepatic apo A-IV mRNA levels after treatment with estrogens (Figure 1). The importance of changes in the clearance of plasma lipids and apolipoproteins is clearly demonstrated by the alterations in LDL-receptor gene expression after treatment with thyroid hormones and estrogens. Both ethinylestradiol (Figure 1) and L-thyroxine (Figure 3) increase hepatic LDL receptor mRNA levels, thereby contributing to the decreased plasma concentrations of the ligands for the LDL-receptor, apo B and apo E. Hypothyroidism induced by n-propylthiouracil, in contrast, decreases liver LDL-receptor mRNA, which results in the increased plasma cholesterol, apo E and apo B concentrations (Figure 2).

However, it is also evident from these studies that the observed alterations in the expression of the examined genes are not always sufficient to explain the changes observed in plasma lipid and apolipoprotein levels. For instance, although all corticosteroids examined have qualitatively similar ef-

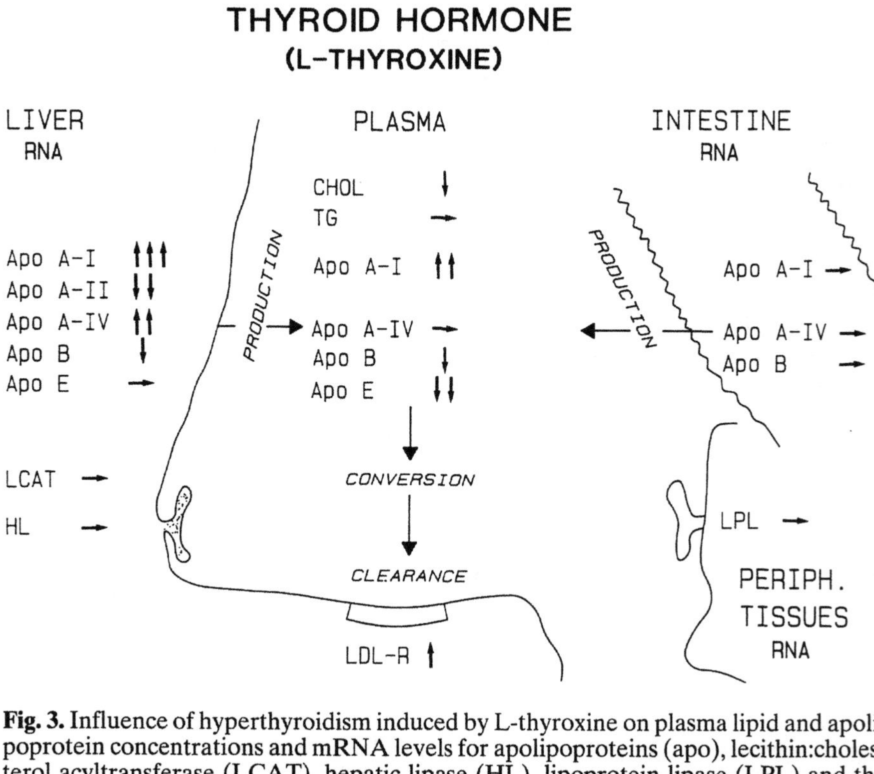

Fig. 3. Influence of hyperthyroidism induced by L-thyroxine on plasma lipid and apolipoprotein concentrations and mRNA levels for apolipoproteins (apo), lecithin:cholesterol acyltransferase (LCAT), hepatic lipase (HL), lipoprotein lipase (LPL) and the LDL-receptor (LDL-R) in rat liver, intestine and adipose tissue. Changes are expressed in a similar fashion as described in figure 1. For more details see Staels et al. 1990a & 1990c

fects on liver apolipoprotein mRNA levels, distinct effects are observed on plasma apo E and apo A-IV concentrations (Staels et al. 1991). In addition, plasma cholesterol and triglycerides are regulated in the opposite direction by these hormones. Alterations in hepatic LDL-receptor gene expression cannot explain the divergent effects on plasma cholesterol levels, since LDL-receptor mRNA levels are not significantly influenced by any of the corticosteroids used.

These discordancies show the complexicity of plasma lipoprotein metabolism. Several different explanations may be invoked. First, it is possible that part of the effects of these hormones are indirect. For instance, the alterations in liver apo A-I and hepatic lipase gene expression after treatment with ethinylestradiol are at least in part mediated by changes in the food intake of the animals caused by this hormone (Staels et al. 1989; Staels et al. 1990b). Somewhat similarly, one might hypothesize that the distinct effects of the different corticosteroids on plasma lipids and apolipoproteins may be related to differences in their effects on the intermediate metabolism of lipids and/or carbohydrates. Second, differences in the expression of proteins which are important in lipoprotein metabolism, but which could not yet be included in the

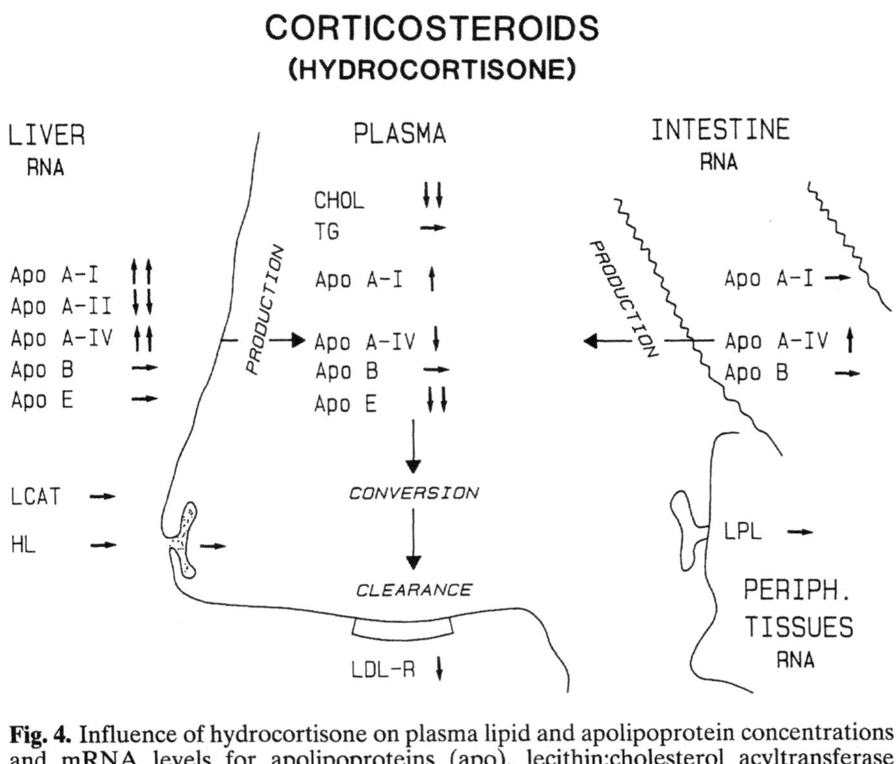

Fig. 4. Influence of hydrocortisone on plasma lipid and apolipoprotein concentrations and mRNA levels for apolipoproteins (apo), lecithin:cholesterol acyltransferase (LCAT), hepatic lipase (HL), lipoprotein lipase (LPL) and the LDL-receptor (LDL-R) in rat liver, intestine and adipose tissue. Changes are expressed in a similar fashion as described in figure 1. For more details see Staels et al. 1990a & 1991

present experiments, may explain the observed discordancies. Indeed, at present little is known on the regulation of the HDL-receptor, the remnant receptor, lecithin:cholesterol acyltransferase or the different cholesterol ester transfer proteins. The isolation and cloning of the corresponding genes should provide us with the opportunity to examine the regulation of these proteins under different conditions. Finally, one should take into account that gene expression can be controlled not only at a transcriptional level, but also at a post-transcriptional level. In the present studies we have mainly considered changes in mRNA concentration, but it is quite conceivable that some of the factors investigated may also influence translation rates and/or protein processing and secretion.

Conclusion

Despite the complex regulation of the expression of these genes, some general conclusions seem to be warranted. First, nearly all the studied genes are regulated in a tissue-specific fashion. Moreover hepatic expression seems to be more tightly controlled than intestinal expression. In the liver apo A-I and apo A-IV gene expression is regulated by steroid and thyroid hormones. In the intestine apo A-I and apo A-IV gene expression is unaffected by the same factors. The total concentration of apo B mRNA in the liver remains relatively constant after treatment with different hormones, but the relative level of apo B-100 and apo B-48 mRNA has been shown to be strongly regulated by thyroid hormones (Davidson et al. 1988). Again this type of regulation is not observed in the intestine. These observations might indicate that a common factor may be involved in the tissue-selective expression and regulation of the apolipoprotein genes.

Second, hepatic apo A-I and apo A-II gene expression is regulated in an opposite direction after treatment with estrogens, thyroid hormones and corticosteroids. This may be important since both apolipoproteins may perform opposite functions in the process of reverse cholesterol transport (Barbaras et al. 1986; Barkia et al. 1991). Accordingly, the opposite regulation of the expression of these genes by hormones and other factors may have important implications for the anti-atherogenic efficiency of HDL particles. Again it would be worthwhile investigating whether a common transcriptional factor is involved in the opposite regulation of these genes.

In conclusion, it is evident from these studies that plasma lipoprotein metabolism is very complex and regulated by a wide variety of factors acting at the level of production and/or catabolism of these lipoproteins and their protein constituents, the apolipoproteins. Further studies on the regulation of the genes involved in lipoprotein metabolism may help us to dissect the molecular mechanisms of primary and secondary forms of hyperlipidemia and ultimately to define strategies to prevent the disastrous consequences of these diseases.

We acknowledge the helpful contributions and discussions with A. van Tol, H. Jansen, G. Ailhaud, J.-C. Fruchart, J. Brunzell, A. Chait in various stages of this work. The excellent technical assistance of J. Rosseels and F. Vanderhoydonc are acknowledged.

References

Barbaras R, Grimaldi P, Négrel R, Ailhaud G (1986) Characterization of high-density lipoprotein binding and cholesterol efflux in cultured mouse adipose cells. Biochem Biophys Acta 888:143–56.

Barkia A, Puchois P, Ghalim N, Torpier G, Barbaras R, Ailhaud G, Fruchart J-C (1991) Differential role of apolipoprotein AI-containing particles in cholesterol efflux from adipose cells. Atherosclerosis 87:135–46.

Chirgwin JM, Przybyla AE, MacDonald RJ, Rutter WJ (1979) Isolation of biologically active ribonucleic acid from sources enriched in ribonuclease. Biochemistry 18: 5294–9.

Dallinga-Thie GM, Groot PHE, van Tol A (1985) Electroimmunoassay of rat apolipoproteins A-I, A-IV and E. A procedure for sample treatment to increase the sensitivity in diluted fractions. J Lipid Res 26:889–92.

Davidson NO, Powell LM, Wallis SC, Scott J (1988) Thyroid hormone modulates the introduction of a stop codon in rat liver apolipoprotein B messenger RNA. J Biol Chem 263:13482–5.

Staels B, Auwerx J, Chan L, van Tol A, Rosseneu M, Verhoeven G (1989) Influence of development, oestrogens and food intake on apolipoprotein A-I, A-II and E mRNA in rat liver and intestine. J Lipid Res 30:1137–45.

Staels B, van Tol A, Verhoeven G, Auwerx J (1990a) Apolipoprotein A-IV messenger ribonucleic acid abundance is regulated in a tissue-specific manner. Endocrinology 126:2153–63.

Staels B, Jansen H, van Tol A, Stahnke G, Will H, Verhoeven G., Auwerx J (1990b) Development, food intake, and ethinylestradiol influence hepatic triglyceride lipase and LDL-receptor mRNA levels in rats. J Lipid Res 31:1211–8.

Staels B, van Tol A, Chan L, Will H, Verhoeven G, Auwerx J (1990c) Alterations in thyroid status modulate apolipoprotein, hepatic triglyceride lipase, and low density lipoprotein receptor in rats. Endocrinology, 27:1144–52.

Staels B, van Tol A, Chan L, Verhoeven G, Auwerx J (1991) Variable effects of different corticosteroids on plasma lipids, apolipoproteins, and hepatic apolipoprotein mRNA levels in rats. Arteriosclerosis, 11:760–9.

Apolipoprotein B mRNA Editing: Modulation and Clinical Implications

D. J. Rader, A. Patterson, T. Eggerman, J. M. Hoeg, and H. B. Brewer, Jr.

Apolipoprotein (apo) B is required for the normal transport of lipids in the plasma and exists in two forms [6]. ApoB-100 is the major apolipoprotein constituent of very low density (VLDL) and low density lipoproteins (LDL). ApoB-48 is the major form of apoB in chylomicrons. ApoB-100-containing LDL are known to be atherogenic lipoproteins, and have a relatively long plasma residence time. In contrast, apoB-48-containing chylomicron remnants are not clearly atherogenic, and are removed from the plasma very rapidly. ApoB-100 and apoB-48 are encoded by the same gene on chromosome 2 [1, 15]. ApoB-100 has a predicted MW of 512 Kd and is a ligand for the LDL receptor [7]. ApoB-48 is approximately half the size of apoB-100 and shares an identical amino terminus with apoB-100, however it lacks the carboxy-terminal domain of apoB-100 [19] which is necessary for LDL receptor binding [7]. The catabolism of apoB-48 is thought to be dependent largely on apoE [13].

Synthesis of ApoB-100 and ApoB-48

The mechanism by which apoB-100 and apoB-48 both arise from the same gene has been elucidated [9, 16, 18, 20]. Analysis of apoB cDNA clones isolated from a human intestinal cDNA library revealed that the intestinal apoB mRNA differed from the genomic DNA by a single base, the result of a substitution of a uridine for a cytidine at nucleotide position 6666. This changes a CAA coding for glutamine to a UAA, which is an in-frame translational stop codon. If no editing of apoB mRNA occurs, the CAA coding for glutamine at residue 2153 remains intact and the mRNA is translated into full-length apoB-100. However, in tissues which have editing capability, such as the human small intestine, codon 2153 is converted to a stop codon and translation is terminated with generation of the truncated apoB-48 protein.

How is it that a single base in the apoB mRNA is singled out for editing? Several laboratories have been actively investigating the sequence requirements for determining the site of editing. In work by Davies, et al. [11], apoB mRNA deletion constructs as short as 26 bases spanning the edited site underwent editing when transfected into and expressed in McArdle 7777 cells, a rat hepatoma cell line. This suggests that as few as 26 nucleotides around the edited site may be sufficient for editing. This region is highly conserved, with 90% homology among human, rabbit, rat and mouse apoB genes [11]. On the other hand, the sequence requirement within the immediate vicinity of the edited nucleotide appears to be rather nonspecific. In work by Chen, et al. [8], mutant constructs with cytidines replacing the adenines in these positions near the editing site resulted in editing of these additional cytidines to uridines. This suggests that the specificity for the edited cytidine at position 6666 lies more in the region around the edited site rather than immediately at the site itself.

ApoB-mRNA-Editing

What is the molecular mechanism of apoB mRNA editing? The simplest mechanism would be a deamination of the cytidine at the 6-position, converting it to a uridine [5, 8]. This type of reaction could be catalyzed by a form of cytidine deaminase. A schematic model of the apoB mRNA editase has been proposed by Smith, et al. [21]. One or more proteins may be responsible for "docking" at the proper site on the apoB mRNA. Relatively short sequences of 12–15 nucleotides on each side of the editing site may be specific sequences responsible for binding this protein or complex. It is speculated that another protein which has cytidine deaminase activity may then bind to the docking complex and be correctly positioned over the cytidine 6666, resulting in deamination of this nucleotide. However, to date no proteins involved in the editing process have been isolated and fully characterized.

The intestine has the ability to edit apoB mRNA and synthesize apoB-48. However, does the human intestine edit its apoB mRNA with 100% efficiency and therefore synthesize only apoB-48, or does some intestinal apoB mRNA remain unedited and therefore become translated into apoB-100? In their analysis of human intestinal apoB biosynthesis, Hoeg and coworkers [17] first demonstrated that human thoracic duct lymph contains some apoB-100. Next, they demonstrated that human intestinal organ cultures synthesize and secrete apoB-100 as well as apoB-48. As further evidence of intestinal apoB-100 synthesis, human intestinal biopsy tissue was stained with a monoclonal antibody specific for apoB-100 and shown to contain a significant amount of apoB-100 within the enterocytes. Finally unedited, or apoB-100, transcripts were found to account for 15% of the human intestinal apoB mRNA. Hence, under fasting conditions, the human small intestine edits an average of 85% of its apoB mRNA transcripts.

Therefore, although the intestine has the ability to edit apoB mRNA and produce apoB-48, not all intestinal apoB mRNA transcripts are edited, resulting in the synthesis and secretion of some apoB-100 by the human small intestine. The physiologic role of these intestinally-derived apoB-100-containing lipoproteins is unknown, but they are likely to have a longer plasma residence time than apoB-48 lipoproteins [13, 22], to be converted to LDL, and to have atherogenic potential. Therefore, modulation of the amount of editing in the small intestine, as well as the potential induction and modulation of editing in the liver, may be clinically important in influencing the development of atherosclerosis.

Influences on the Editing-Process

What is currently known about modulation of the apoB mRNA editing process? In contrast to the human, the rat liver secretes both apoB-100 and apoB-48, with 40–50% of the apoB mRNA in edited form [23]. The rat liver has therefore been utilized to study modulation of the apoB mRNA editing process. Fasting reduces rat hepatic apoB mRNA editing and apoB-48 production by greater than 50%, while refeeding with a glucose-rich diet results in a marked increase in apoB mRNA editing and a doubling of apoB-48 synthesis [12, 3]. In searching for other potential modulators of apoB mRNA editing, Davidson, et al. [10] made rats hypothyroid by feeding them a diet containing propothiouracil for 3–4 weeks. Rats were then made either euthyroid by physiologic hormone replacement, or hyperthyroid by pharmacologic doses of T3, and apoB protein and mRNA were analyzed. ApoB-48 synthesis was affected only slightly by the T_3, with a slight increase in animals which were treated for 7 days. In contrast, apoB-100 production fell rapidly after the T_3 bolus, and in the 7-day treated animals there remained virtually no apoB-100 synthesis. Analysis of hepatic apoB mRNA revealed that the hypothyroid rats had less apoB mRNA editing than the euthyroid rats, while the hyperthyroid rats had significantly increased amounts of editing approaching those in control rat intestine. These results suggest that thyroid hormone may stimulate the apoB mRNA editing process.

What relevance does this have for human pathophysiology? Hyperthyroidism is associated with decreased total cholesterol, LDL, and apoB levels, whereas hypothyroidism is associated with increased apoB and apoB-containing lipoproteins [14]. The fractional catabolic rates of both VLDL and LDL are increased in hyperthyroidism and decreased in hypothyroidism [1, 2]. This is felt to be due, at least in part, to modulation of LDL receptor expression. However, if thyroid hormone can affect the human liver and/or intestine and stimulate more apoB mRNA editing, then more apoB-48 particles and less apoB-100 particles would be produced, resulting in an overall faster rate of VLDL catabolism. Conversely, hypothyroidism is often associated with overproduction of VLDL, especially in obese patients [1]. It is possible that this could result from diminished intestinal editase activity with subsequent overproduction of apoB-100 VLDL particles. Importantly, treatment of the underlying thyroid condition usually normalizes the plasma concentrations and kinetics of the apoB-containing lipoproteins [14].

Summary

In summary, apoB mRNA editing is a complex and highly regulated process. In the human, the intestine synthesizes and secretes lipoproteins containing apoB-100 as well as lipoproteins containing apoB-48, whereas the liver secretes only apoB-100 containing lipoproteins under normal physiologic conditions. ApoB-48 particles are rapidly cleared from the plasma and in most circumstances are unlikely to play a significant role in atherogenesis. Conversely, apoB-100 particles are likely to be converted to LDL and remain in the plasma for a longer period of time, with the potential of participating in the atherosclerotic process. The modulation of apoB mRNA editing may therefore be an important determinant of plasma apoB levels and hence of atherogenic risk.

References

1. Abrams JJ, Grundy SM, Ginsberg H (1981) Metabolism of plasma triglycerides in hypothyroidism and hyperthyroidism in man. J Lipid Res 22:307–22.
2. Abrams JJ, Grundy SM (1981) Cholesterol metabolism in hypothyroidism and hyperthyroidism in man. J Lipid Res 22:323–38.
3. Baum CL, Teng B, Davidson N (1990) Differential regulation of hepatic apolipoprotein gene expression and apoB-100 mRNA editing following alterations in hepatic lipogenesis. Clin Res 38:482A.
4. Blackhart BD, Ludwig EM, Pierotti VR, et al. (1986) Structure of the human apolipoprotein B gene. J Biol Chem261:15364–15367.
5. Bostrom K, Lauer SJ, Poksay KS, Garcia Z, Taylor JM, Innerarity TL (1989) Apolipoprotein B48 RNA editing in chimeric apolipoprotein EB mRNA. J Biol Chem 264:15701–15708.
6. Brewer HB,Jr, Gregg RE, Hoeg JM, Fojo SS (1988) Apolipoproteins and lipoproteins in human plasma: an overview. Clin Chem 34:B4–B8.
7. Brown MS, Goldstein JL (1986) A receptor-mediated pathway for cholesterol homeostasis. Science 232:34–47.
8. Chen SH, Li XX, Liao WS, Wu JH, Chan L (1990) RNA editing of apolipoprotein B mRNA. Sequence specificity determined by in vitro coupled transcription editing. J Biol Chem 265:6811–6816.
9. Chen SH, Habib G, Yang CY, et al. (1987)Apolipoprotein B-48 is the product of a messenger RNA with an organ-specific in-frame stop codon. Science 238: 363–366.
10. Davidson NO, Powell LM, Wallis SC, Scott J (1988) Thyroid hormone modulates the introduction of a stop codon in rat liver apolipoprotein B messenger RNA. J Biol Chem 263:13482–13485.
11. Davies MS, Wallis SC, Driscoll DM, et al. (1989) Sequence requirements for apolipoprotein B RNA editing in transfected rat hepatoma cells. J Biol Chem 264: 13395–13398.
12. Davis RA, Boogaerts JR, Borchardt RA, Malone-McNeal M, Archambault-Schexnayder J (1985) Intrahepatic assembly of very low density lipoproteins. Varied synthetic response of individual apolipoproteins to fasting. J Biol Chem 1985; 260:14137–14144.
13. Gregg RE, Brewer HB, Jr (1988) The role of apolipoprotein E and lipoprotein receptors in modulating the in vivo metabolism of apolipoprotein B-containing lipoproteins in humans. Clin Chem 34:B28–B32.

14. Heimberg M, Olubadewo JO, Wilcox HG (1985) Plasma lipoproteins and regulation of hepatic metabolism of fatty acids in altered thyroid states. Endocr Rev 6:590–607.
15. Higuchi K, Monge JC, Lee N, et al. (1987) The human apoB-100 gene: apoB-100 is encoded by a single copy gene in the human genome. Biochem Biophys Res Commun 144:1332–1339.
16. Higuchi K, Hospattankar AV, Law SW, Meglin N, Cortright J, Brewer HB, Jr (1988) Human apolipoprotein B (apoB) mRNA: identification of two distinct apoB mRNAs, an mRNA with the apoB-100 sequence and an apoB mRNA containing a premature in-frame translational stop codon, in both liver and intestine. Proc Natl Acad Sci USA 85:1772–1776.
17. Hoeg JM, Sviridov DD, Tennyson GE, et al. (1987) Both apolipoproteins apoB-100 and apoB-48 are synthesized by the human intestine. J Lipid Res 31:1761–1769.
18. Hospattankar AV, Higuchi K, Law SW, Meglin N, Brewer HB, Jr (1987) Identification of a novel in-frame translational stop codon in human intestine apoB mRNA. Biochem Biophys Res Commun 148:279–285.
19. Marcel YL, Innerarity TL, Spilman C, Mahley RW, Protter M, Milne RW (1987) Mapping of human apolipoprotein B antigenic determinants. Arteriosclerosis 7:166–175.
20. Powell LM, Wallis SC, Pease RJ, Edwards YH, Knott TJ, Scott J (1987) A novel form of tissue-specific RNA processing produces apolipoprotein-B48 in intestine. Cell 50:831–840.
21. Smith HC, Kuo SR, Backus JW, Harris SG, Sparks CE, Sparks JD (1991) In vitro apolipoprotein B mRNA editing: Identification of a 27S editing complex. Proc Natl Acad Sci USA 88:1489–1493.
22. Sparks CE, Rader DJ, Marsh JB (1983) Metabolism of two forms of apolipoprotein B of VLDL by rat liver. J Lipid Res 24:156–166.
23. Tennyson GE, Sabatos CA, Higuchi K, Meglin N, Brewer HB, Jr (1989) Expression of apolipoprotein B mRNAs encoding higher- and lower-molecular weight isoproteins in rat liver and intestine. Proc. Natl Acad Sci USA 86:500–504.

Hormonal Regulation of Lipoprotein Metabolism in Obesity

W. O. RICHTER, A. C. SOENNICHSEN, M. M. RITTER, P. SCHWANDT

Severe obesity leads to an increased risk for coronary heart disease. Although the relative increase of risk is low (from 1.0 in normal weight subjects to 1.55 in persons with an overweight >40%) in comparison to other diseases (e. g. increase in mortality rate for diabetes mellitus from 1.0 to 7.55 in obese women) it has great importance for public health [1]. The factors leading to the increased risk include a higher frequency of hypertension and other cardiovascular risk factors. A positive energy balance and especially a high fat intake during weight gain negatively influences serum lipoproteins. But also under a weight maintaining diet distinct lipoprotein alterations are present in obesity. Characteristic are an increase of serum triglycerides, VLDL-triglycerides and VLDL-cholesterol, and decreased HDL-cholesterol concentrations. Table 1 illustrates these findings in 158 men also indicating that there is no correlation between BMI and total or LDL-cholesterol. In moderate obesity there may be a slight increase of total cholesterol but in severe obesity (BMI >36 kg/m^2) often low cholesterol concentrations can be observed (Fig. 1).

In obesity also typical alterations in the plasma concentration of hormones which play a role in lipoprotein metabolism can be found. Under basal and stimulated conditions (oral glucose tolerance test) higher concentrations of insulin, C-peptide and glucagon can be observed as shown for women in table 2. But no correlation to insulin receptor binding (Table 2) could be obtained. This finding has also been shown in a very carefully done study by

Table 1. Linear regression analysis between body-mass-index (BMI) and lipoproteins in healthy men (n = 158)
(BMI range from 20–50 kg/m^2)

	BMI
cholesterol	0.10
LDL-cholesterol	0.11
VLDL-cholesterol	0.20*
Apolipoprotein B	0.21*
triglycerides	0.31***
HDL-cholesterol	0.22**
VLDL-triglycerides	0.34***
Apolipoprotein A I	0.16

* = p<0.05; ** = p<0.01; *** = p<0.001

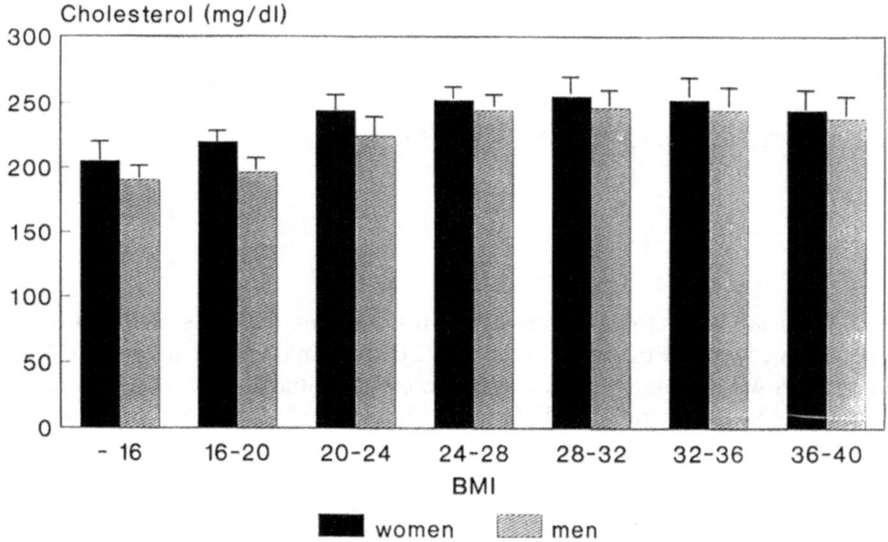

Fig. 1. Correlation between Body-Mass-Index and total cholesterol in 75477 persons (Bavarian Cholesterol Screening Project)

Bolinder et al. [2]. Furthermore there was no correlation to other gastrointestinal hormones like pancreatic polypeptide, β-endorphin or somatostatin. The two hormones which showed a significant correlation to the extent of overweight were insulin and glucagon. This is emphasized by the fact that after an oral glucose load the area under the curve of insulin, C-peptide or glucagon was significantly greater in obese premenopausal women (Table 3). Yet, when the basal or stimulated levels of hormones (after 100 g of oral glucose) were analyzed together with serum lipoproteins no significant correlations could be found (Table 4). Only pancreatic polypeptide showed a significant correlation but on the one hand the physiological function of this peptide is unclear and on the other plasma levels do not correlate to BMI. So it can be stated that there is no correlation between basal or stimulated hormone levels which are elevated in obesity and serum lipoproteins and also not to the extent of insulin binding to its receptor.

Table 2. Linear regression analysis between BMI and basal levels of hormones in lean and obese premenopausal women (n = 42)

	BMI
insulin	0.36*
insulin receptor binding	0.25
pancreatic polypeptide	0.18
glucagon	0.31*
β-endorphin	0.12
somatostatin	0.01

* = $p<0.05$

Table 3. Area under the curve after an oral glucose load (100 g) in lean and obese premenopausal women

	lean n=12 mean BMI= 21,4 kg/m^2	SEM	obese n=30 mean BMI= 36,4 kg/m^2	SEM
insulin (μU·h)	133	18	252***	47
pancreatic polypeptide (pg·h)	152	26	157	20
c-peptide (nmol·h)	12.2	1.4	17.9*	2.1
glucagon (pg·h)	34.7	9	158***	34
β-endorphin (pmol·h)	7.9	2.3	5.0	3.0

* = $p<0.05$; *** = $p<0.001$

Table 4. Linear regression analysis between basal levels of hormones and the area under the curve after an oral glucose load (100 g) and lipoproteins in obese men (n = 40, mean age 49±2 years, BMI between 28 and 40 kg/m^2)

	chol.	LDL-c.	triglyc.	HDL-c.
insulin	−0.08	−0.07	0.03	−0.10
insulin area	−0.17	−0.24	0.13	−0.02
pancreatic polypept. (PP)	0.30*	0.38**	−0.29*	0.23
PP-area	0.27*	0.34*	−0.23	016
β-endorphin	0.05	0.07	0.09	−0.08
β-endorphin area	0.11	0.13	0.00	0.13

* = $p<0.05$; ** = $p<0.01$

Lipoprotein Abnormalities

Independently from the extent of overweight the distribution of body fat seems to play an important role regarding cardiovascular risk and lipoprotein abnormalities. Already in 1947 Vague observed that the risk for the development of atherosclerosis was much higher in "android" than in "gynoid" obesity [3, 4]. The extent of "android" and "gynoid" adipose tissue distribution can be measured by the waist-to-hip-ratio (waist circumference determined at the umbilicus and hip circumference at the trochanter major). A large number of epidemiologic studies has been carried out to prove Vague's hypothesis [5–10]. Table 5 shows that there is a strong correlation of the WHR to total and LDL-cholesterol, apolipoprotein B but also to triglyceride rich lipoproteins.

But it is noteworthy that also a strong correlation of the waist-to-hip ratio with the BMI and also with age has been observed. As both these factors influence serum lipoprotein concentration only a multivariant analysis can establish which factor is primarily important for these lipoprotein alterations. In men as well as in women multivariate analysis showed (Table 6) the primary dependency of triglycerides, VLDL-cholesterol and VLDL-triglycerides on adipose tissue distribution, of HDL-cholesterol on BMI and of total

Table 5. Linear regression analysis between adipose tissue distribution (WHR) and age, body-mass-index as well as serum lipoproteins in men (n = 158)

	WHR
age	0.44***
cholesterol	0.32***
LDL-cholesterol	0.24**
VLDL-cholesterol	0.37***
VLDL-triglycerides	0.44***
Apolipoprotein B	0.50***
BMI	0.53***
triglycerides	0.40***
HDL-cholesterol	−0.09
Apolipoprotein A I	−0.05

** = $p<0.01$; *** = $p<0.001$

and LDL-cholesterol on age. How can it be explained that both obesity per se and the type of adipose tissue distribution are accompanied with different lipoprotein alterations?

The Role of Insulin

Insulin serum levels do not only depend on the extent of overweight, as shown above, but also on adipose tissue distribution (WHR). Higher insulin

Table 6. Stepwise multiple linear regression between age, WHR, BMI and lipoproteins in men (n = 158)

dependent variable	step	independent variable	mult. R	FEV %	p
cholesterol	1	age	0.37	14	<0.001
	2	WHR	0.43	18	<0.01
LDL-cholesterol	1	age	0.36	13	<0.001
triglycerides	1	WHR	0.43	18	<0.001
HDL-cholesterol	1	BMI	0.17	3	<0.05
VLDL-cholesterol	1	WHR	0.44	19	<0.001
VLDL-triglycerides	1	WHR	0.46	21	<0.001

in women (n = 147)

dependent variable	step	independent variable	mult. R	FEV %	p
cholesterol	1	age	0.49	23	<0.001
	2	BMI	0.51	25	<0.05
LDL-cholesterol	1	age	0.44	19	<0.001
triglycerides	1	WHR	0.36	13	<0.001
HDL-cholesterol	1	BMI	0.39	14	<0.001
VLDL-cholesterol	1	WHR	0.20	3	<0.05
VLDL-triglycerides	1	WHR	0.37	13	<0.001

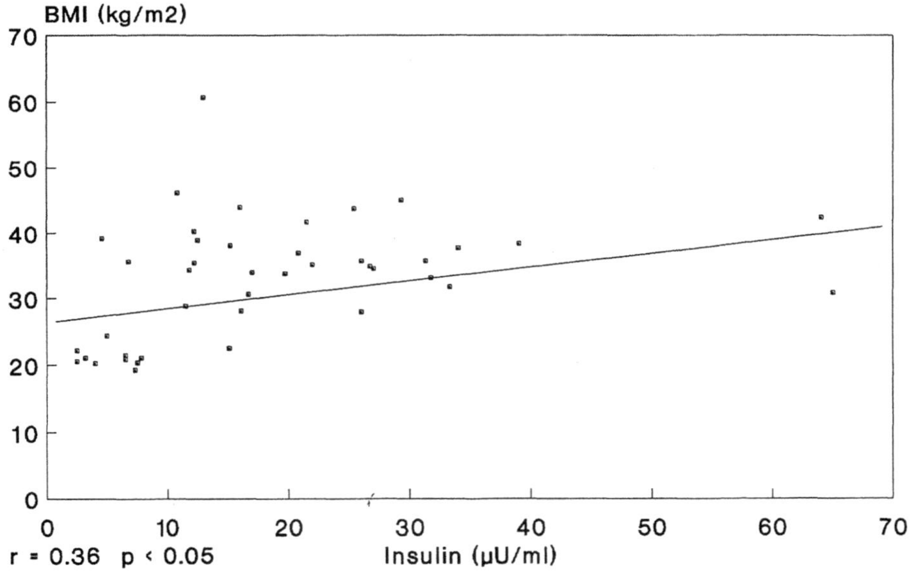

Fig. 2. Correlation between Body-Mass-Index and insulin in normal and overweight premenopausal women (n = 42)

levels are found in "android" obesity. So obesity and/or a high WHR can be associated with hyperinsulinism. But when only basal insulin levels are regarded several obese people have normal values (Fig. 2), but after stimulation e.g. under hyperglycemic glucose clamping (elevation of blood glucose by 125 mg/dl within 15 minutes and maintenance of the new blood glucose level for 2 hours) a "postprandial" hyperinsulinism can be observed (Fig. 3). This hyperinsulinism can also be shown after a mixed meal or other stimuli [11] and is accompanied by a peripheral insulin resistance. It is not definitely clarified whether insulin resistance at the cellular level or hyperinsulinism due to β-cell hypersecretion is the primary defect because peripheral insulin resistance leads to hyperinsulinsm and on the other hand even in normal subjects insulin resistance can be induced by insulin infusion within 24 hours [12]. Factors which are involved in cellular insulin resistance are known. There is no altered binding of insulin to the receptor (which has also been shown in our patients (Table 2)) as suggested several years ago.

The structure of the insulin receptor has been elucidated recently [13, 14]. After binding of insulin to the alpha-subunit the insulin receptor acts as a tyrosine kinase. In the β-subunit tyrosine residues are phosphorylated. In states with insulin resistance (e.g. obesity, type II diabetes) the extent of this so-called autophosphorylation is decreased [15]. As autophosphorylation is the prerequisite for the metabolic activity of insulin the consequence is a reduced glucose transport and a diminished antilipolytic effect. Furthermore the insulin response of glucose transporters (proteins which are responsible for the intracellular transport of glucose) to move from the cellular pool to the cell

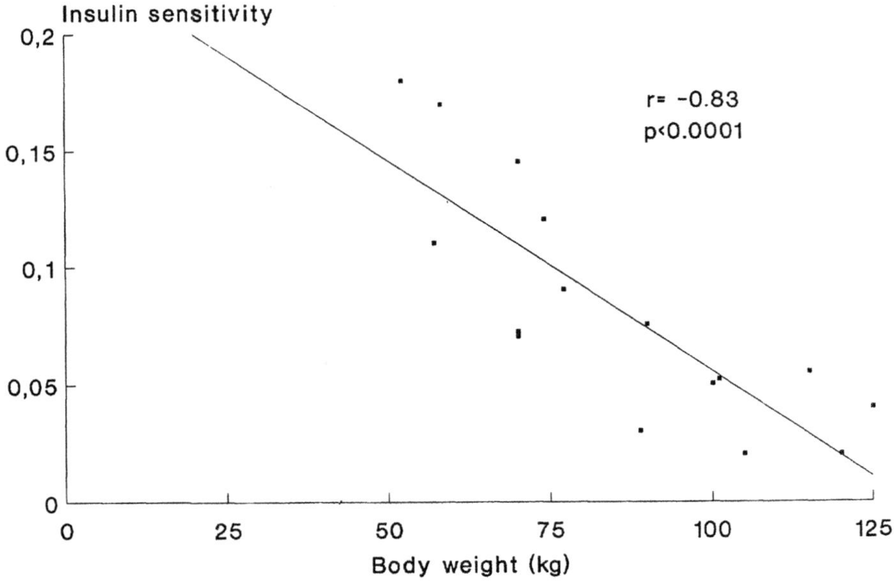

Fig. 3. Correlation between body weight and insulin sensitivity $\left(\frac{\text{mg} \times \text{ml}}{\text{kg} \times \text{min} \times \text{mU}}\right)$ (n=16). Insulin sensitivity established by hyperglycemic glucose clamping

membrane is lower and also the absolute number in the pool [16]. As there is no absolute correlation between insulin sensitivity and hyperinsulinism interindividually, insulin resistance has to be evaluated by other means. We determined insulin sensitivity by hyperglycemic clamping in 16 lean and obese subjects. There was a highly significant correlation between insulin sensitivity and the extent of overweight (Fig. 3) and also to adipose distribution (WHR, $r = 0.53$). When serum lipoproteins were included into the analysis there was a highly significant correlation to VLDL triglycerides, VLDL cholesterol and HDL-cholesterol in men and women (data for women shown in Table 7). So the extent of insulin resistance is very important for the regulation of serum lipoprotein levels in obesity.

Pathophysiologic Mechanisms in Obesity

The pathomechanism linking alterations in serum lipoproteins and insulin sensitivity may be the following: insulin resistance leads to an accelerated release of free fatty acids (decreased antilipolytic activity of insulin) from adipose tissue preferably from the abdominal site ("android obesity"). The increased flow of free fatty acids to the liver results in a stimulation of VLDL secretion.

The lower correlation of HDL-cholesterol and insulin sensitivity and the primary dependency of HDL-cholesterol on the extent of overweight may be

Table 7. Linear regression between insulin sensitivity and lipoproteins in women (n = 8). Insulin sensitivity established by hyperglycemic clamping

	M/I	M/C
BMI	−0.777	−0.703
WHR	−0.690	−0.44
triglycerides	−0.868	−0.660
VLCL-C	−0.798	−0.636
VLDL-TG	−0.653	−0.452
HDL-C	+0.51	+0.41

$$M / I = \frac{mg \times ml}{kg \times min \times \mu U}$$

$$M / C = \frac{mg \times l}{kg \times min \times nmol}$$

explained by an additional factor. HDL-binding to adipocytes is increased in obesity [17], furthermore there may be a shift of cholesteryl ester and apolipoprotein AI from HDL to triglyceride rich lipoproteins [18].

Besides insulin sensitivity other factors may play a role: β-lipotropin and β-endorphin are produced in a common precursor hormone, the proopiomelanocorticotropin (POMC). Especially β-lipotropin is able to stimulate lipolysis from adipose tissue of laboratory animals and man in near physiological concentrations [19, 20]. As an increased flux of free fatty acids may be responsible for the accelerated VLDL secretion from the liver an increased serum concentration of these peptides could be a causative factor. Indeed Solerte et al. [21] were able to show a correlation between plasma levels of these hormones and serum triglyceride concentration. Some authors found the plasma concentrations of these hormones to be elevated in subtypes of obesity [22, 23].

As insulin sensitivity appears to be one of the causes of these probably atherogenic lipoprotein alterations, factors which improve the action of insulin are important in the treatment.

A hypocaloric diet is able to normalize triglyceride rich lipoproteins within a few days but also leads to a further decrease of HDL-cholesterol. The increase of HDL cholesterol due to weight reduction becomes evident after reaching steady state conditions regarding body weight and nutrition. An increased physical activity with eucaloric nutrition can improve insulin sensitivity [24] and also lipoprotein metabolism.

References

1. Garfinkel L (1986) Overweight and cancer. Ann Intern Med 1985; 103:1034–6.
2. Bolinder J, Lithell H, Skarfors, E, Arner, P. (1986) Effects of obesity, hyperinsulinema, and glucose intolerance on insulin action in adipose tissue of sixty-year-old men. Diabetes 35:282–290.
3. Vague J (1947) La differenciation sexuelle, facteur determinant des formes de l'obesite. Presse Med 55:339–340.

4. Vague J (1956) The degree of masculine differentiation of obesities: a factor determining preposition to diabetes, atherosclerosis, gout and uric calculous disease. Am J Clin Nutr 4:20.
5. Haffner SM, Fong D, Hazuda HP, Pugh JA, Patterson JK (1988) Hyperinsulinemia, upper body adiposity, and cardiovascular risk factors in non-diabetics. Metabolism 37:338–345.
6. Kissebah AH, Vydelingum N, Murray R (1982) Relation of body fat distribution to metabolic complications of obesity. J Clin Endocrinol Metab 54:254–260.
7. Hartz AH, Rupley DC, Rimm AA (1984) The association of girth measurements with disease in 32856 women. Am J Epidemiol 119:71–80.
8. Evans DJ, Hoffmann RG, Kalkhoff RK, Kissebah AH (1984) Relationship of body fat topography to insulin sensitivity and metabolic profiles in premenopausal women. Metabolism 33:68–75.
9. Ostlund RE, Staten M, Kohrt WM, Schultz J, Malley M (1990) The ratio of waist-to-hip circumference, plasma insulin level, and glucose intolerance as independent predictors of the HDL2 cholesterol level in older adults. New Engl J Med 322:229–234.
10. Donahue RP, Abbott RD, Bloom E, Reed DM, Yano K (1987) Central obesity and coronary heart disease in men. Lancet 1:821–823.
11. Polonsky KS, Given BD, Van Cauter E (1988) Twenty-four-hour profiles and pulsatile patterns of insulin secretion in normal and obese subjects. J Clin Invest 81: 442–448.
12. Marangou AG, Weber KM, Boston RC, Aitken PM, Heggie JCP, Kirsner RLG, Best JD, Alford FP (1986) Metabolic consequences of prolonged hyperinsulinemia in humans. Diabetes 35:1383–1389.
13. Kahn CR, White MF (1988) The insulin receptor and the molecular mechanism of insulin action. J Clin Invest 82: 1151–1156.
14. Morgan, DO, Roth, RA (1987) Acute insulin action requires insulin receptor kinase activity: introduction of an inhibitory monoclonal antibody into mammalian cells blocks the rapid effects of insulin. Proc Nat Acad Sci USA 84:41–45.
15. Takayama S, Kahn CR, Kubo K, Foley JE (1988) Alterations in insulin receptor autophosphorylation in insulin resistance: correlation with altered sensitivity to glucose transport and antilipolysis to insulin. J Clin Endocr Metab 66:992–999.
16. Garvey WT, Huecksteadt, TP, Matthaei S, Olefsky JM (1988) Role of glucose transporters in the cellular insulin resistance of type II non-insulin-dependent diabetes mellitus. J Clin Invest 81:1528–1536.
17. Grundy SM, Denke MA (1990) Dietary influences on serum lipids and lipoproteins. J Lipid Res 31:1149–1172.
18. Jimenez JG, Fong BS, Julien P, Despres JP, Angel A (1988) Weight loss in massive obesity: reciprocal changes in plasma HDL cholesterol and HDL binding to human adipocyte plasma membranes. Metabolism 37:580–586.
19. Richter WO, Schwandt P (1985) Peptide hormones and lipolysis in rabbit adipocytes. Horm Metabol Res 17:127–130.
20. Richter WO, Schwandt P (1985) Physiologic concentrations of β-lipotropin stimulate lipolysis in rabbit adipocytes. Metabolism 34:539–543.
21. Solerte SB, Fioravanti M, Petraglia F, Facchinetti F, Patti AL, Schifino N, Genazzani AR, Ferrari E (1988) Plasma beta endorphin, free fatty acids and blood lipid changes in type 2 (non-insulin dependent) diabetic patients. J Endocrinol Invest 11:37–42.
22. Facchinetti F, Giovannini C, Barletta C, Petraglia F, Buzzetti R, Burla F, Lazzari R, Genazzani AR, Scavo D (1986) Hyperendorphinemia in obesity and relationships to affective state. Physiol Behav 36:937–940.
23. Givens JR, Wiedemann E, Andersen RN, Kitabchi AE (1980) β-endorphin and β-lipotropin plasma levels in hirsute women: correlation with body weight. J Clin Endocrinol Metab 50:975–976.
24. DeFronzo RA, Sherwin RS, Kraemer N (1987) Effect of physical training on insulin action in obesity. Diabetes 36:379–1385.

Lipoproteins in Patients with NIDDM: Effects of Different Therapeutic Approaches

J. SCHNEIDER

Before considering effects of lipid-lowering drugs it may be useful to summarize the essentials in the pathogenesis of hyperlipoproteinemia in subjects with NIDDM. According to the work of Reaven and his group, an increase in VLDL secretion occurs with decreasing insulin response, the second fact is that actual serum levels of triglycerides (TG) are higly determined by VLDL secretion. The increase in VLDL production seems to depend mainly from a high turnover of free fatty acids [1].

Another abnormality may be even more important in terms of atherogeneity: the composition of the particles separated as VLDL is altered to a cholesterol enriched lipoprotein relatively poor in TG. As it has been shown by Kissebah and others, the newly synthezised particles are of normal composition [2]. Changes in the composition of lipoproteins isolated from plasma therefore seem to reflect altered catabolism. Indeed, VLDL in NIDDM are converted to LDL in a lower degree and the resulting LDL are TG-enriched and contain less esterified cholesterol. These alterations may contribute to the increased cardiovascular risk in NIDDM.

We have studied the effect of two drugs known to lower lipids in NIDDM which are less frequently used than the standard treatment with fibrates.

1. Biguanide Study

This study was performed as a randomized placebo controlled two-arm-trial in 40 patients with NIDDM and hyperlipoproteinemia. Patients with unstable angina pectoris, myocardial infarction within one year, myocardial and respiratory insufficiency, impaired renal function, hepatic disease besides fatty infiltration, gastrointestinal disorders, malignancies and psychological abnormalities were excluded as well as women of childbearing potential. Inclusion was independent of the quality of glucose control. Diagnosis of type II diabetes was clinical. In doubtful cases, it was confirmed by positive insulin and C-peptide response after glucagon stimulation. All patients had been treated effectively with diet and sulfonylurea derivatives for at least 6 months. Hyperlipoproteinemia was defined as total cholesterol >250 mg/dl and/or triglycerides >200 mg/dl. The patients were aged between 37 and 77 years, all of them were overweight, with a body mass index (weight (kg)/height (m^2)) from 25 to 32.

Methods

At 0, 4, 8 and 12 weeks, fasting blood samples were taken and VLDL, LDL, HDL_2 and HDL isolated by preparative ultracentrifugation in 50-3 Ti rotors in L50 centrifuges (Beckman Palo Alto). Cholesterol, triglycerides and phospholipids in serum and in the density fractions were determined using standard enzymatic methods; HDL-cholesterol was also determined after polyanion precipitation of apo B-containing lipoprotein (Boehringer, Mannheim). Apo-A-I, A-II and B were determined in sérum and density fractions using radial immunodiffusion on standard commercial plates (Immuno, Heidelberg). HbA1 was determined using a microcolumn method (Biorad, Munich). Compositions were calculated according to the formula (Chol. + TG + PL + Apo B) mg/dl = 100 %. Statistical comparison of the data of the patients who completed the 12-week trial (n = 18 in the metformin group, n = 16 in the placebo group) was made by the Wilcoxon matched-pairs signed-rank test (comparison within groups) and the Mann-Whitney test (comparisons between groups).

Results

Lipid and Apoprotein Concentrations

With metformin there was a significant reduction in triglycerides in serum and in the density fractions VLDL and LDL. There was a simultaneous, highly significant fall in cholesterol in VLDL, which produced a significant fall in total cholesterol even though LDL-cholesterol rose by 12 % (n.s.). Cholesterol concentrations in the HDL- and HDL_2-UC-fractions were unchanged; the increase in (precipitated) HDL-cholesterol was not significant. The following changes in apoproteins were significant: an increase in apo A-I and apo A-II in the HDL-fraction after 12 weeks, and a fall in apo B in the VLDL fraction after 12 weeks. By contrast, apart from a slight increase in A-I in the HDL_2-fraction and in HDL phospholipids, no significant changes were observed in the control group (Table 1).

Lipoprotein Composition

The relative content of cholesterol and phospholipids in the VLDL fell by 14 and 21 % and triglycerides increased by 9 %. In the LDL fraction there was an increase in the proportion of cholesterol of 8 % and a fall in triglycerides of 23 %, the latter being highly significant (Table 2). There were no changes in composition in the placebo group.

Table 1. Lipids and apoproteins in serum and lipoprotein classes in the metformin group
Values are mean ±SD in mg/dl

		Weeks: 0	4	8	12
Serum	Chol	242± 43	232± 43*	232± 44**	233± 51*
	TG	340±207	250±161**	259±210***	242±156***
	PL	305± 59	284± 59	284± 39	290± 45
	A I	140± 27	142± 29	151± 29	151± 40
	A II	55± 10	55± 10	56± 10	53± 10
	B	104± 16	107± 20	105± 17	98± 20
VLDL	Chol	56± 43	38± 29*	35± 20***	34± 27***
	TG	243±195	173±156***	186±205***	170±144***
	PL	68± 60	40± 32	40± 28*	38± 28*
	B	15± 10	9± 5	13± 9	10± 4*
LDL	Chol	133± 42	140± 37	145± 41	147± 39
	TG	56± 24	51± 15	49± 17	45± 16
	PL	96± 23	96± 27	106± 30	107± 27*
	B	72± 23	72± 27	76± 21	70± 20
HDL_2	Chol	24± 15	22± 15	23± 14	22± 15
	TG	11± 4	10± 3	9± 3	10± 3
	PL	41± 28	40± 22	38± 12	37± 21
	A I	30± 20	27± 22	35± 24	37± 17
	A II	14± 5	17± 5	14± 6	14± 5
HDL	Chol	47± 22	46± 19	48± 19	49± 21
	TG	24± 8	21± 5	19± 5	19± 5
	PL	89± 35	81± 30	83± 19	84± 28
	A I	98± 24	98± 37	106± 35	110± 29*
	A II	39± 9	38± 12	42± 8	44± 7*
HDL-Chol (precip.)		38± 17	40± 18	41± 18	42± 20

* P>0.05; ** P>0.01; ***P>0.001 compared to initial value

Table 2. Effect of treatment with metformin for 12 weeks on composition of VLDL and LDL in hyperlipoproteinemie NIDDM (n=18)
Values are mean ±SD in m g/dl

		Weeks: 0	4	8	12
VLDL	Chol	14.7±2.1	16.1±2.1	12.7±3.4	12.7±3.6
	TG	*62.9±6.5	64.4±7.8	66.8±2.7	68.3±5.2
	PL	18.7±6.1	15.5±5,4*	15.2±4.1	14.7±2.9*
	Apo B	3.7±1.5	4.0±1.8	5.2±5.5	4.3±2.5
LDL	Chol	36.6±5.1	38.9±5.1	38.4±3.1	39.5±3.5**
	TG	16.6±6.9	14.7±5.0	13.4±4.7	12.7±4.9***
	PL	27.2±4.0	26.7±4.1	28.0±2.2	28.8±2.1
	Apo B	20.0±4.6	19.7±4.4	20.2±3.7	19.0±3.6*

* P>0.05; ** P>0.01; ***P>0.001 compared to initial value

Relationship with Improvement of Diabetic Control

In the group treated with metformin there was a highly significant fall in HbA1c from 6.76 to 6.08, in the control group the value fell from 6.48 to 6.34 (n.s.).

Relationship between HbA1c and changes in all lipoprotein parameters were computed and expressed as Spearman correlation coefficients. Out of the parameters with significantly different changes between the groups, only the increase in the relative content of cholesterol in VLDL revealed a significant correlation with the improvement in HbA1c.

2. HMG-CoA-Reductase Inhibitor Study

Patients with stable NIDDM, treated with diet and, if necessary, with oral glucose-lowering drugs, who were in our 4–8 weekly control for at least one year and who showed persistent hyperlipoproteinemia, were instructed as to the purpose, possible risks and benefits of the planned open study. No patient with secondary hyperlipoproteinemia due to any other cause than diabetes mellitus was included, nor were patients with triglycerides greater than 500 mg/dl and phenotypes I and V. After having given informed consent, according to the declaration of Helsinki, the patients entered the study. All of them were instructed in the use of a standard antidiabetic diet which included the criteria of the cholesterol lowering diet according to American Heart Association (AHA) stage I. Eventuel lipid lowering drug treatment was stopped for at least four weeks. Concomitant drugs potentially exerting lipid changes such as beta-blockers and diuretics were exchanged for lipid neutral compounds like calcium-antagonists or angiotensin converting-enzyme inhibitors or, at least, were kept constant in dosage throughout the study. After a placebo period of four weeks 20 patients presented a total cholesterol value exceeding 250 mg/dl. They received simvastatin 10 mg/d, which was titrated up to 20 or 40 mg/d so long as total cholesterol exceeded 205 mg/dl, for a total of six four-weeks periods. The final dose was 10 mg in four, 20 mg in six and 40 mg/d in ten patients.

Methods

Aliquots of serum were submitted to preparative ultracentrifugation to isolate VLDL, IDL, LDL and $HDL_{2/3}$ at densities of 1.006, 1.019, 1.063, 1.125 and 1.210 according to standard procedures. In the obtained fractions Chol, H, TG and PL were determined, as well as Apo B in VLDL, IDL and LDL and Apo A-I in HDL_2 and HDL_3, The lipid constituents in the serum and in the subfractions were determined by enzymatic colour tests (Boehringer Mannheim), Apo B and A-I by kinetic nephelometry (Beckman). Statistical analysis was performed by Student's paired t-test.

Results

During the study, there was no change in the glucose control of the patients: The initial mean value for HbA1c was 7.51; the final value 7.68, with an increase in nine, a decrease in 6 and a constant value in five cases. These changes were not significant. Significant decreases occured in total, VLDL- and LDL-cholesterol and -triglycerides (Table 3). The fall in total phospholipids originates mainly in VLDL. No significant differences were observed in the lipid components of IDL.

In both HDL-fractions the TG-content was reduced, the fall reached significance in HDL_3. A shift from HDL_3 to HDL_2 was seen in PL and Apo-I.

Apoprotein B decreased by 30% in total. LDL contributed more than VLDL in absolute terms, but the relative decrease was more pronounced in VLDL than in LDL. Like the lipid constituents the Apo B content in IDL remained unchanged. HDL-cholesterol measured by polyanionic precipitation showed not differences (41.3 ± 3.6 initially, 40.8 ± 3.1 mg/dl after 24 weeks of treatment). In the fractions with major changes of the absolute values, VLDL and LDL, no change occured in the composition of VLDL. In the LDL there was at the very end of the study a significant increase in PL at the expense of TG and cholesterol, while the cholesterol/Apo B ratios remained unchanged (initially 1.64, finally 1.58, n.s.).

3. Discussion

Treatment with the biguanide resulted in a decrease of triglycerides in the VLDL fraction accompanied by a reduction in VLDL Apo B, the only non-exchangeable and therefore consistent component of the individual particle in the lipolytic cascade. This leads to the conclusion that there is a reduction in the number of circulating VLDL particles. Monitoring the composition of the VLDL fraction, however, shows that the relative amount of Apo B in the VLDL total mass increased in the course of treatment, i.e. this is not a purely numerical reduction in an unchanged particle spectrum. This is particularly clear if one considers the changes in the composition of lipid in the VLDL-fraction.

The relative increase in triglycerides and the relative reductions in cholesterol and phopholipids in the course of treatment in our study suggest a shift away from the lipoproteins of the catabolic final stage in this density fraction towards particles richer in triglycerides and hence normalization. By contrast to VLDL, in the LDL fraction the proportion of triglycerides fell and cholesterol rose, with a constant apo B content. An increased proportion of triglycerides in the LDL fraction has been reported in type II diabetes with a certain amount of agreement. Multidispersity of LDL and equivocal changes in composition of VLDL and LDL as a characteristic of NIDDM independent from the VLDL-TG levels has been described recently. As VLDL and LDL in circulation constitute a continuous spectrum of precursor-product particles

Table 3. Cholesterol, triglycerides, phospholipids and apoprot ein A I and B in serum and lipoprotein density classes in mg/dl during treatment with simvastatin

		Weeks: 0	4	8	12	24
Serum	Chol	288±10	226±21***	225±10***	212± 8***	208± 9***
	TG	385±72	283±4**	245±39**	248±50***	230±37***
	Phos	299±14	246±12***	262± 5**	243±11***	239±11***
	Apo A I	153± 7	169± 7*	156± 6	159± 7	152± 8
	Apo B	139± 8	108± 7***	113± 7***	104± 7***	102± 6***
VLDL	Chol	67±12	52± 9*	46± 8**	43± 9***	38± 7***
	TG	340±67	260±43*	219±42**	224±52***	195±39***
	PL	87±17	60±10*	50± 9**	53±11**	60±10**
	Apo B	22± 7	19± 3	17± 2*	16± 3*	14± 2**
IDL	Chol	17± 2	14± 2*	17± 3	14± 2	14± 2
	TG	16± 2	16± 2	19± 2	18± 2	21± 2*
	Phos	13± 2	12± 1	14± 3	12± 1	13± 2
	Apo B	9± 1	9± 1	11± 2	10± 1	9± 1
LDL	Chol	176±13	142±10***	142± 8***	131± 7***	129± 9***
	TG	44± 3	34± 2***	37± 3***	35± 3**	29± 2
	Phos	99± 9	79± 6**	85± 5*	79± 4*	90± 9
	Apo B	107± 7	82± 6***	86± 4***	89± 5*	82± 5***
HDL_2	Chol	26± 3	25± 2	24± 3	28± 3	26± 3
	TG	13± 1	12± 2	11± 1	12± 2	12± 1
	PL	35± 5	33± 3	35± 4	39± 4	49± 5***
	Apo A I	35± 5	35± 4	38± 6	42± 5*	44± 5***
HDL_3	Chol	30± 1	19± 1	19± 1*	17± 1*	18± 1
	TG	10± 1	9± 1	9± 1	9± 1	9± 1*
	PL	35± 2	31± 2	33± 2	30± 2	27± 1*
	Apo A I	65± 4	66± 4	58± 3*	61± 4	55± 3**

* P>0.05; ** P>0.01; ***P>0.001 compared to initial value by t-test

with a constant Apo B content but with varying amounts of other apoprotein components and lipid composition (large triglyceride rich VLDL, remnants, IDL, LDL) we observed a dilution of the "middle population", this means a reduction of the NIDDM-prone multidispersity. As for HDL, the elevated proportion of triglycerides in HDL affecting both HDL_2 and HDL_3, seems to arise through the free exchange of components between triglyceride-rich lipoproteins and HDL, and recedes with the reduction of hypertriglyceridemia.

The question arises whether the changes in lipoprotein particle numbers and compositions observed in this study are biguanide specific. Hypocaloric diet treatment of type II diabetes also reduces hypertriglyceridemia and increases HDL-cholesterol if 10-12 % weight loss is achieved [3]. Under sulfonylurea therapy, low HDL values were more likely to be observed, or HDL-related parameters (HDL-cholesterol, Apo A-I) did not increase [4]. Under insulin treatment, both, reductions in triglycerides and increases in HDL-cholesterol were observed, IDL-cholesterol was unchanged [5]. In the only available report on insulin-induced changes with detailed data on lipoprotein composition in all density classes, a 60 % reduction in VLDL-triglycerides was reported, and in contrast to our results, a fall in the proportion of triglycerides and an increase in cholesterol and phospholipids in VLDL, IDL and LDL were isolated separately: their ratio to one another did not change substantially [6]. The differences between insulin – and metformin – mediated changes are in favour of a metformin-specific effect on VLDL and LDL, which is confirmed by our finding that the changes in VLDL and LDL composition did not correlate significantly with the improvement of HbA1c. This view is further strengthened by the fact that the triglyceride lowering effect of biguanides has been observed in non-diabetic subjects as well as in diabetics [7].

Any decrease in LDL concentration in NIDDM deserves special consideration: it has been shown that higher degrees of LDL-glycolysation interfere with the high-affinity binding to LDL-receptors [8]. As no significant change occured in the HbA1c values during the study with simvastatin, this unspecific effect can be ruled out in the former discussion of our results. The considerable decrease in total serum cholesterol from 288.2 to 208.2 mg/dl reaches the degree which may have been expected from the trials with simvastatin in primary hypercholesterolemias. Different and noteworthy is the contribution of VLDL-cholesterol which decreased by nearly 50 % compared to the fall in LDL-cholesterol by 27 % only. Due to the proportional reduction of the components, the composition of VLDL and the Chol/Apo B ratio in the LDL remained unchanged. Just the opposite has been described recently during treatment of heterozygous familial hypercholesterolemia with lovastatin, namely a reduction in total mass, but in addition a modification in the composition of VLDL and LDL: the decrease in VLDL confined to the $VLDL_3$ fraction and the decrease in the relative content of cholesterol in the LDL has been discussed as the result of increased catabolism (increased clearance of VLDL-remnants, preferential removal of cholesterol-enriched

LDL, or reduced residence time of LDL particles), while the decrease in LDL without changes in composition observed earlier in moderate primary hypercholesterolemia could be explained more by a reduction in the input rate of LDL-Apo B [9, 10, 11]. Reduced input instead of increased removal as a mode of action of HMG-CoA-reductase inhibitors has been suggested in human cholesterol storage disease [12]. This interpretation of our own results in compositions of VLDL and LDL is strenghened by the fact that IDL remained absolutely unchanged.

Although an assumption which remains to be proven by turnover studies, this mode of action suggests itself in dyslipidemia of NIDDM, a condition characterized primarily by overproduction of VLDL and not by decreased LDL-receptor activity as in heterozygous familial hypercholesterolemia.

References

1. Reaven GM (1990) Pathogenesis of Abnormal Very-Low-Density and High-Density Lipoprotein Metabolism in Patients with Non-Insulin-Dependent Diabetes Mellitus. In: Belfiore F, Molinatti GM, Reaven GM (eds) Tissue-Specific Metabolic Alterations in Diabetes. Front Diabetes, 10:134–144
2. Kissebah AH, Alfarsi S, Evans DJ, Adams PW (1982) Integrated regulation of very-low-density lipoprotein triglyceride and apolipoprotein B kinetics in non-insulin-dependent diabete mellitus. Diabetes 31:217–225
3. Kennedy L, Walshe K, Hadden DR, Weaver JA, Buchanan KD (1982) The effect of intensity dietary therapy on serum high-density lipoprotein cholesterol in patients with type II (non-insulin-dependent) diabetes mellitus: A prospective study. Diabetologia 23:24–27
4. Bar-On H, I andau D, Berry E (1977) Serum high-density lipoprotein and university group diabetes program results. Lancet I:761
5. Hughes TA, Clements RS, Fairclough PK, Bell DSH, Segrest JP (987) Effect of insulin therapy on lipoproteins in non-insulin-dependent diabetes mellitus (NIDDM) Atherosclerosis 67:105–114
6. Taskinen MR, Kuusi T, Helve E, Nikkilä EA, Yki-Järvinen H (1988) Insulin therapy induces antiatherogenic changes of serum lipoproteins in non-insulin-dependent diabetes. Arterosclerosis 8:168–177
7. Sirtori CR, Tremoli E, Sirtori M, Conti F, Paoletti R (1977) Treatment of hyertriglyceridemia with metformin. Atherosclerosis 26:583–592
8. Kesaniemi YA, Witztum JL, Steinbrecher UP (1983) Receptor-mediated cataboism of low density lipoprotein in man. Quantitation using glycosylated low density lipoprotein. J Clin Invest 71:950–959
9. Nozaki S,. Vega GL, Haddox RJ, Dolan ET, Grundy SM (1990) Influence of lovastatin on concentrations and composition of lipoprotein subfractions. Atherosclerosis 84:101–110
10. Grundy SM, Vega GL (1985) Influence of mevinolin on metabolism of low density lipoproteins in primary moderate hypercholesterolemia. J Lip Res 26: 1464–1475
11. Vega GL, Krauss RM, Grundy SM (1990) Pravastatin therapy in primary moderate hypercholesterolaemia; changes in metabolism of apolipoprotein B-containing lipoprotiens. J Int Med 227:81–94
12. Ginsberg HN, Le N-A, Short MP, Ramakrishnan R, Desnick RJ (1987) Suppression of apolipoprotein B production during treatment of cholesterol ester storage disease with lovastatin: implication for regulation of apolipoprotein B. J Clin Invest 80:1692–1698

Postprandial Course of Triglycerides and the Atherogenic Lipoprotein Particle "IDL" in Type 2 Diabetics with Normal Fasting Values

P. WEBER, S. AUSIEKER, R. PROBST, U. KRAUSE, W. PRELLWITZ, J. BEYER, and J. SCHREZENMEIR

Introduction

Hypertriglyceridemia, which represents the main lipid disorder in patients with type2 diabetes mellitus [8], is supposed to be one of the factors contributing to the increased prevalence of atherosclerosis in these patients. In the Paris Prospective Study, hypertriglyceridemia was found to be an independent risk factor for coronary heart disease in normocholesterolemic diabetic patients [3]. More recently, these epidemiological data were confirmed by Bergstrom et al. [1]. They described an independent association of coronary heart disease with elevated fasting total triglycerides and with an impaired glucose tolerance in 219 second-generation Japanese-American men.

Triglyceride (TG) levels are mainly influenced by nutritional factors [9]: human individuals eating regular meals exhibit a prolonged postprandial elevation of triglyceride levels most part of the day. However, most former studies on lipid disorders in diabetic patients employed fasting TG values. So far, less is known about the postprandial course of atherogenic lipoprotein particles in these patients. In addition, there is increasing evidence that an altered lipoprotein composition may exist in type2 diabetics even with normal fasting TG values. Klein et al. [6] found an abnormal apoprotein composition of VLDL and Kasama et al. [4] reported on increased cholesterol concentrations of the LDL in type2 diabetics with normolipidemic fasting values. The aim of the present study was to evaluate the postprandial course of the TG, the triglyceride-rich lipoproteins and in particular of the atherogenic IDL fraction following the ingestion of a standardized lipid load in type2 diabetics with normal fasting TG.

Methods

10 patients with type2 diabetes mellitus (mean age: 57 ± 11 years) and 13 healthy subjects (mean age: 26 ± 2 years) participated at the study. For the diabetics, mean body mass index (BMI) was 29.8 ± 6.7 kg/m^2. HbAI values had an average of $9.4 \pm 1.8\%$ (range: 6.2–12.5%), duration of diabetes was 10 ± 7 years. The healthy controls were nonsmokers, were free of diseases, were not on any medication and they had a normal body weight (BMI: 19.1 ± 1.6 kg/m^2).

After an overnight fast subjects were given a standardized oral lipid load at 8 a.m. The formula contained 1017 kcal consisting of 30 g protein (12 kcal %), 58 g fat (51 kcal %), 75 g carbohydrates (30 kcal %), 10 g alcohol (7 kcal %) and 600 mg cholesterol. After the meal, subjects fasted for 24 h with free excess to water. During the whole test period they stayed hospitalized. Blood samples (25 ml) were drawn via a small indwelling catheter prior to the test meal and at 4, 5, 9, 11, 14 and 24 h after the meal. We measured in each sample: triglycerides in the whole serum, the triglyceride-rich lipoprotein fraction and the IDL concentrations. The triglyceride-rich lipoprotein fraction (chylomicrons, VLDL and IDL) was separated in a preparative ultracentrifuge (Beckmann L 3–50) with a 50 Ti rotor at a density of 1.019 g/ml (24 h, 4 °C, 40000 rpm). We determined IDL in this fraction by lipid electrophoresis. Triglyceride concentrations were assessed by the "Triglycerides fully enzymatic UV-Test" (Merck, Darmstadt, Germany) [2]. Lipid electrophoresis was carried out by the "Lipidophor all in 12 kit" (Immuno, Heidelberg, Germany) [10].

Statistical analysis was done by the programme SAS 6.03. Differences were analysed for significance using paired student t-test or Wilcoxon rank test when values did not have a normal distribution.

Results

All subjects had normal fasting triglyceride values: total triglyceride levels were 88±27 mg/dl in healthy subjects and 141±47 mg/dl in patients with type2 diabetes mellitus. After ingestion of the lipid load total triglycerides rose significantly ($p<0.001$). The initial triglyceride response of diabetic patients was very similar to that in healthy subjects. The postchallenge triglyceride peak occured in all subjects at 4-5 h postprandially. In contrast, the decline of the triglycerides was remarkably delayed in the type2 diabetics (Table 1). The postprandial nadir occured in the diabetic patients significantly later at 13 h, whereas the control group reached its minimum at 9 h ($p<0.01$). The triglycerides in the triglyceride-rich lipoprotein fraction showed a similar postprandial course and again, the nadir was significantly delayed in the diabetes group ($p<0.001$). The postprandial decrease of the total triglycerides after the postchallenge peak was 19 mg/dl · h in the diabetics and 34 mg/dl · h in the healthy subjects ($p<0.07$). In the triglyceride-rich lipoprotein fraction we observed a decline of 13 mg/dl · h in the diabetics versus 22 mg/dl · h in the control group ($p<0.02$)(Fig. 1).

The patients with type 2 diabetes mellitus tended to have higher fasting IDL concentrations than nondiabetics: 24±10 mg/dl in the diabetics versus 16±10 mg/dl. Following the ingestion of the lipid load, peak values for IDL in nondiabetics were observed at 5 h (12±5 mg/dl). The diabetics, by contrast had the highest IDL values at 9 h (26±15 mg/dl). At this point in time IDL levels were significantly ($p<0.05$) higher in the diabetes group and tended to stay elevated much longer (24 h) than in nondiabetics.

Table 1. Triglyceride levels in whole serum and in the triglyceride-rich lipoprotein fraction (TGL) before and after a standardized oral lipid load in 10 patients with type2 diabetes mellitus and in 13 healthy controls. Values are depicted as mean ±SD. (pp = postprandial)

	basal	5 h pp	9 h pp	11 h pp	14 h pp	24 h pp
	triglycerides (mg/dl)					
Total:						
diabetes	141±47[a]	241±70	164±64[c]	123±46[c]	100±57[a]	147.±50
healthy	88±37	206±78	71±26	68±24	77±27	96±30
TGL:						
diabetes	80±38[a]	138±56	85±44[c]	69±40[b]	37±31	75±36
healthy	47±24	117±74	30±16	29±14	36±18	48±22

[a] = $p<0.05$; [b] = $p<0.01$; [c] = $p<0.005$

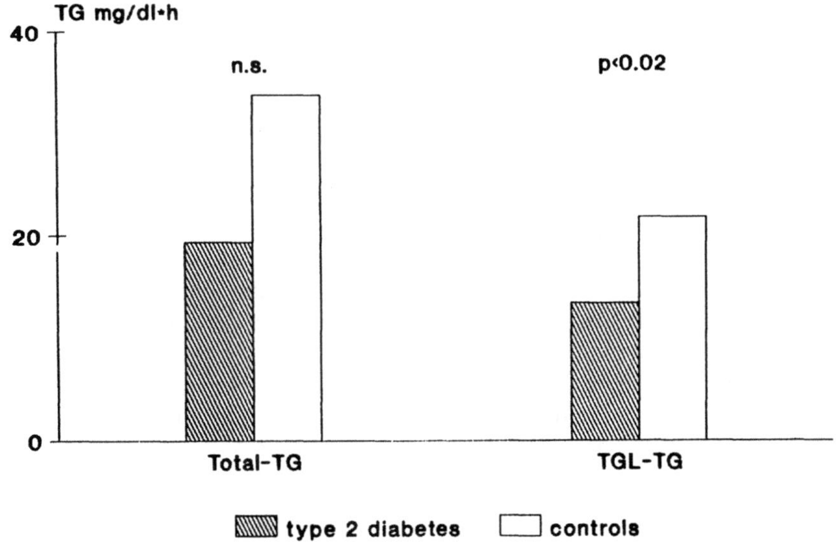

Fig. 1. Postprandial decrease of the triglycerides (TG) (mean value from timepoint 5 h to 9 h postprandially) in the whole serum and in the triglyceride-rich lipoprotein fraction (TGL) in 10 patients with type2 diabetes mellitus

Discussion

In the past, numerous studies had been carried out on the subject of lipoprotein metabolism in diabetics. They all had emphasized that hypertriglyceridemia is a frequent disorder in patients with type2 diabetes mellitus. However, most of these previous studies had measured fasting triglyceride levels only.

Reports on postprandial lipoprotein metabolism in type2 diabetes mellitus are rare and even controversial.

Some years ago, Nikkila et al [7] had applied an intravenous fat tolerance test in diabetics and found that the fractional elimination rate of exogenous fat was not different from that of nondiabetics with identical fasting triglyceride levels. However, the intravenous route for application of fat may not be a suitable approach for studying postprandial lipoprotein metabolism. More recently, in type2 diabetics with slightly elevated fasting triglycerides Taskinen et al. [12] described both an increased production and a decreased clearance of VLDL as indicated by a lowered fractional catabolic rate after a so-called "standard" breakfast. On the other hand, Katsilambros [5], who also applied a "standard" breakfast as a lipid load failed to demonstrate any differences in the postprandial course of triglycerides in type2 diabetics as compared to nondiabetics.

The studies cited above, however, had several disadvantages: they varied in the way of application, in the fat content of the load and with regard to the postprandial observation period. In the present study we used a manufactured, standardized oral lipid load containing identical ingredients for the whole series of the study.

Under these conditions, type2 diabetics who had completely normal fasting triglyceride values had a significantly prolonged postprandial hypertriglyceridemia as compared to healthy controls. As judged by our preliminary data, triglycerides reached their peak values in both diabetics and controls at about 5 h. The subsequent decline of the triglycerides was significantly slower in the diabetics. This could be demonstrated for both total triglycerides and for triglycerides in the triglyceride-rich lipoprotein fraction. This finding favours the thesis of a defective clearance rather than an increased production of triglycerides in diabetic patients. Furthermore, our data clearly indicate that the postprandial observation period in the study of Katsilambros [5] was too short to detect any difference between these two groups.

So far, it remains unclear whether this delayed clearance of the postprandial triglycerides has any impact on the development of atherosclerosis in these patients. In the present study, however, we found in patients with type2 diabetes mellitus a significantly higher and longer postprandial elevation of the IDL, which are well known for their atherogenic potential. Interestingly, in support of this interpretation it was recently reported that patients with coronary heart disease also have a prolonged postprandial hypertriglyceridemia [11].

In summary, our data provide evidence that in patients with type2 diabetes mellitus a prolonged postprandial hypertriglyceridemia may exist even when fasting values are normal. Thus, the actual state of the triglyceride metabolism in diabetics can not be judged exactly by determining fasting values only.

References

1. Bergstrom RW, Leonetti DL, Newell-Morris LL, Shuman WP, Wahl PW, Fujimoto WY (1990) Association of plasma triglyceride and C-peptide with coronary heart disease in Japanese-American men with a high prevalence of glucose intolerance. Diabetologia 33:489–96.
2. Bucolo G, David H (1973) Quantitative determination of serum triglycerides by the use of enzymes. Clin Chem 19:476–82.
3. Cambien F, Jacqueson A, Richard JL, Warnet LM, Ducimetiere P, Claude JR (1986) Is the level of serum triglyceride a significant predictor of coronary death in "normocholesterolemic" subjects?. Am J Epidemiol 124:626–32.
4. Kasama T, Yoshino G, Iwatani I et al. (1987) Increased cholesterol concentration in intermediate density lipoprotein fraction of normolipidemic non-insulin-dependent diabetes. Atherosclerosis 63:263–6.
5. Katsilambros N, Stavrou P, Manjaris V et al. (1978) Serum Lipids in Diabetics and Nondiabetics after a Test Meal. Diabetologia 15:244.
6. Klein RL, Lyons TJ, Lopes-Virella MF (1990) Metabolism of very low- and low-density lipoproteins isolated from normolipidaemic Type 2 (non-insulin-dependent) diabetic patients by human monocyte-derived macrophages. Diabetologia 33:299–305.
7. Nikkila AE, Kekki M (1973) Plasma Triglyceride Transport Kinetics in Diabetes Mellitus. Metabolism 22:1–22.
8. Reaven GM (1987) Abnormal Lipoprotein Metabolism in Non-Insulin-Dependent Diabetes Mellitus. Amm J Med 83 (suppl 3A):31–40.
9. Schlierf G, Oster P, Seidel D, Raltzer H, Schellenberg B, Heuck CC, Wicklein RL (1975) Tagesprofile von Plasmalipiden und Lipoproteinen bei Patienten mit endogener Hypertriglyceridamie (Typ IV Hyperlipoproteinamie). Klin Wschr 55: 161–7.
10. Seidel D, Wieland H, Ruppert C (1973) Improved Techniques for assessment of plasma lipoprotein pattern. I Precipitation in gel after electrophoresis with polyanionic compouds. Clin Chem 19:737–9.
11. Simpson HS, Williamson CM, Olivecrona T et al. (1990) Postprandial lipemia, fenofibrate and coronary artery disease. Atherosclerosis 85:193–202.
12. Taskinen MR, Beltz WF, Harper RM et al. (1986)Effects of NIDDM on Very-Low-Density Lipoprotein Triglycerides and Apolipoprotein B Metabolism. Diabetes 35:1268–77.

Differences Between Basal and Postprandial Circadian Variation of Insulin Sensitivity in Healthy Subjects and Type 1 Diabetics

J. SCHREZENMEIR, F. TATÒ, S. TATÒ, C. LAUE, and J. BEYER

Summary

The circadian variation of basal and postprandial plasma glucose and insulin levels was analyzed in 10 healthy individuals and 10 type-1 diabetic patients treated with a glucose-controlled insulin infusion system. In both groups the postprandial glycemic response to identical low caloric mixed meals ingested at 8 a.m., 1 p.m., and 6 p.m. increased significantly in the course of the day. In contrast to findings in conventional insulin therapy mean postprandial insulin requirements of the diabetic patients increased from 8.5 ± 3.0 IU for breakfast to 9.1 ± 2.0 for lunch and 10.9 ± 3.0 for dinner ($p<0.01$). Repeated studies with different test meals showed that in the diabetic patients the circadian deterioration of carbohydrate tolerance was reduced after a fiber rich meal with low glycemic effect and insulin requirements. In the healthy subjects basal insulin levels at 8 a.m. were 35% higher than at 1 and 6 p.m. ($p<0.01$). Basal insulin requirements of the diabetic patients increased significantly from a night value of 0.7 IU/h to 1.18 IU/h during the early morning and remained constant throughout day time.

We conclude that circadian changes in postprandial carbohydrate tolerance are independent from the endogenous rhythmics of basal glucose metabolism. In diabetic patients the circadian pattern of postprandial responses is substantially determined by exogenous factors like diet composition, caloric intake and therapeutic regimen.

Introduction

In non-diabetics glucose tolerance was found to decrease during the course of the day [1–7]. This physiological diurnal fluctuation in glucose tolerance is suggested to be due to enhanced postprandial insulin release in the morning after overnight fasting [3, 2, 5–9] and to reduced peripheral insulin sensitivity in the evening [5, 6, 10, 11].

In contrast to these findings in non diabetic humans the postprandial insulin requirements of insulin dependent diabetic patients observed during intensified conventional insulin therapy and continuous subcutaneous insulin infusion are usually highest after breakfast, and glucose tolerance tends to improve as the day goes on [12–17] This circadian pattern was confirmed by data resulting from computer assisted insulin dosage adjustment [18]. According to this empirical finding it is a well established rule in insulin therapy to calculate a higher insulin dose per gram carbohydrate for breakfast than for the later meals [19, 20].

Although these diurnal changes in insulin need are generally accepted only very few studies have been designed to investigate systematically the influence of the time of the day on the postprandial glucose metabolism of diabetic patients. It remains widely unclear which mechanisms are responsible for the observed circadian variation, and why diabetic patients seem to have an inverse diurnal rhythm compared to healthy humans.

Circadian changes in cortisol, growth hormone and catecholamines have been discussed as possible explanations of increased basal insulin requirements during the early morning (dawn phenomenon) [21–25] and could also contribute to a reduced glucose tolerance after breakfast.

It is remarkable, however, that several studies performed in type 1 diabetic patients with strictly standardized test meals and/or intravenous insulin delivery systems failed to confirm highest insulin requirements per gram carbohydrate for breakfast [3, 26, 30]. This finding suggests that in type-1 diabetes factors like the composition of the diet, the size and sequence of the meals and the unphysiological subcutaneous insulin substitution may decisively influence the observed diurnal pattern of insulin requirements and possibly mask an endogenous circadian rhythm of glucose tolerance.

The aim of this study was to analyze the circadian variation of the basal and postprandial insulin requirements of 10 type-1 diabetic patients ingesting identical test meals at intervals of five hours, using a glucose controlled insulin infusion system. The influence of the composition of the meal on the diurnal changes in insulin requirements was studied by investigating each subject on three separate days with three different standardized mixed meals. The results of the diabetic patients are compared to the corresponding blood glucose and insulin levels of 10 healthy volunteers.

Subjects, Materials and Methods

Subjects

10 type-1 (insulin-dependent) diabetic patients and 10 healthy volunteers were studied. The type-1 diabetic patients were 8 males and 2 females aged between 19 and 25 years. Their mean body mass index was 21.6 ± 1.49 kg/m^2 ranging from 19.3 to 24.3 kg/m^2. In all patients a relevant endogenous insulin secretion was excluded by radioimmunoassay of C-peptide before and one hour after ingestion of a test meal. Besides diabetes, the subjects suffered from no other organic disease, had no clinical signs of neuropathy and no drugs other than insulin were taken. The duration of diabetes was 2 to 8 years (mean 4.9 ± 2.2 years). The daily insulin dose before the study ranged from 36 to 66 U (mean 51 ± 11 U/day).

The healthy volunteers were 4 males and 6 females aged between 20 and 32 years (mean 24.6 ± 3.5 years), with a mean body mass index of 22.1 ± 3.4 kg/m² (range 18.9 to 31.0 kg/m²).

The aim of the study was explained to each participant, and all gave their informed consent. The study was conducted in accordance to the principles of the Declaration of Helsinki.

Test Meals

The studies were performed with three standardized test meals. The three test meals were designed to be of equal energy (1366 kJ = 326 kcal) and carbohydrate content (36 g according to Souci et al. [31]), while differing in the content of protein, fat, and dietary fibre. The composition of the test meals, which were called "continental breakfast", "English breakfast" and "milk", is shown in Table 1. In a previous study these test meals were shown to differ greatly in their glycemic effect [32]. The glycemic index in healthy subjects

Table 1. Composition of the test meals (according to Souci et al. [31]). PR = protein, COH = carbohydrate

1. Continental Breakfast	PR (g)	FAT (g)	COH (g)	kcal	kJ
50 g roll	3.4	0.3	28.7	139	562
20 g butter	0.1	16.6	0.1	155	649
25 g diabetics' jam	–	–	7.5	30	126
	3.5	16.9	36.3	324	1357

2. English Breakfast	PR (g)	FAT (g)	COH (g)	kcal	kJ
200 g kidney beans (can)	13.2	1.0	33.2	186	778
15 g onions	0.2	–	1.5	6	25
50 g egg (fried)	6.4	5.6	0.4	83	347
8 g bacon	0.7	5.2	–	53	220
	20.5	11.8	35.1	328	1370

3. Milk	PR (g)	FAT (g)	COH (g)	kcal	kJ
750 ml skimmed milk (0.3% fat)	26.3	0.8	36.0	263	1095
20 ml cream (30% fat)	–	6.3	0.7	63	264
	26.3	7.1	36.7	326	1359

(compared to 36 g of glucose) was 76% for the continental breakfast, 54% for the English breakfast and 23% for milk.

Additional enzymatic food analysis (according to Holm et al. [33]) and to Methods of enzymatic food analysis, Boehringer Mannheim, Biochemica, FRG, 1984) was performed to verify the carbohydrate content of the test meals. The analyzed carbohydrate content was 34.2 g in the continental breakfast, 33.5 g in the English breakfast, and 36.1 g in the milk. The dietary fibre content of the English breakfast was analyzed according to Asp [34] and amounted to 12.6 g (i.e. 15.9 g/100 g).

Experimental Design

On each study day the same test meal was ingested at 8.00 a.m., 1.00 p.m. and 6.00 p.m. during 10 minutes. No other food was allowed between the meals. For all subjects each study day closed at 10 p.m. Every subject was studied on three separate days and on each day one of the three different test meals was applied in random order.

The experiments with type-1 diabetic patients were performed under glucose feedback control using a Biostator (GCIIS, Bayer Diagnostics, München, FRG). The constants used were: KR = 80, KF = 45, BI = 100, QI = 30, RI = 15, FI = 400, BD = 60, QD = 30, RD = 40, FD = 200. These constants have been shown to achieve normoglycemia with physiological postprandial blood glucose fluctuations [35].

On the previous day the diabetic patients were instructed to inject only the regular insulin portion of their evening dose and at 10 p.m. the patients were admitted to the clinic and connected to the Biostator. The healthy volunteers were admitted on the study day after overnight fast of at least 12 hours.

In the healthy volunteers venous blood samples were collected via an indwelling catheter in an antecubital vein 5 minutes before and 15, 30, 45, 60, 90, 120, 180, 240, 300 minutes after the start of each meal. Blood was collected in precooled tubes containing 1000 KIE Aprotinin (Trasylol, Bayer, Leverkusen, FRG) and 2 mm EDTA per ml, immediately centrifugated at 4 °C, and the plasma stored at –20°C for later analysis.

Laboratory Methods

Plasma free insulin was measured in all samples by radioimmunoassay as described elsewhere [36]. Antibody-bound insulin was separated by precipitation with polyethyleneglycol [37]. A commercially available radioimmunoassay kit was used for measurement of plasma C-peptide (Byk-Sangtec Diagnostica, Dietzenbach). Plasma glucose in the samples from healthy individuals was measured by the glucose oxydase method.

Data Analysis

In type-1 diabetic patients the insulin infusion rate per minute and the protocol of the continuous blood glucose monitoring were provided by the Biostator. Meal-related insulin requirements were calculated for a period of 4 hours after ingestion by subtracting the individual basal insulin requirements from the total amount of insulin infused during each postprandial period. The basal insulin requirements subtracted were obtained for each study day from the insulin infusion rate between 0.00 and 6.00 a.m.

The insulin response of the healthy subjects was quantified by integrating the areas under the postprandial plasma insulin curves for a period of 4 hours after ingestion. As for insulin requirements of diabetic patients preprandial plasma insulin values were subtracted to obtain meal-related postprandial data. The kinetics of the postprandial changes was analyzed by determining the maximum of the curve, the time point of the maximum and the period for the curve to return to the fasting value.

The dawn phenomenon was assessed in type-1 diabetic patients by comparing basal blood glucose and insulin requirements during the periods 0.00 to 3.00 a.m., and 5.00 to 8.00 a.m. Preprandial basal blood glucose and insulin requirements were established in the diabetic patients from 7.30 to 8.00 a.m., 12.30 a.m. to 1.00 p.m., and 5.30 to 6.00 p.m.

Defining insulin sensitivity by the glucose decreasing action of insulin and index was calculated which was defined by 1000/(I basal)(GL basal) for the basal state and 100/(I max)(GL max) for the postprandial state. (I basal = basal plasma insulin level [mU/l] in non-diabetics and basal insulin requirements [U/l] in diabetics; GL basal = basal blood glucose level [mmol]; I max = maximal postprandial insulin level (healthy subjects) or requirements (diabetics); GL max = maximal postprandial glucose level).

The results are reported as mean ± standard deviation (SD). The analysis was performed for all data with and without discrimination between the three different test meals. For latter calculations all 30 study days (3 per person) of each group of subjects were pooled.

Statistical significance of differences was assessed using Student's paired t-test. Since all pairwise comparisons of 3 means were performed, a difference was declared significant, if the corresponding p-value was less than 0.01.

Results

1. Healthy Subjects. For all test meals preprandial plasma glucose and insulin levels were higher at 8.00 a.m. than at 1.00 and 6.00 p.m. (Table 2). Averaged for all tests performed mean 8.00 a.m. plasma glucose was 6.1% and plasma insulin 38.6% higher than at 1.00 p.m. ($p<0.01$). Accordingly the insulin sensitivity index at 8.00 a.m. was significantly lower than at 1.00 and 6.00 p.m. ($p<0.01$). During the course of the day the smallest glycemic response was

Table 2. Preprandial plasma glucose, insulin levels and insulin sensitivity index (1000/insulin · glucose level) of 10 healthy individuals for the three test meals ("continental breakfast", "English breakfast", "milk") and for all meals pooled.

Test Meal Time	Plasma Glucose mmol/l	Plasma Insulin μU/ml	Insulin Sensitivity Index l^2/mmol · mU
Continental			
8.00 a.m.	5.09±0.39	8.86±4.31	27.5±12.5
1.00 p.m.	4.64±0.37 *	4.60±1.89 **	55.2±22.0 **
6.00 p.m.	4.82±0.50 $	5.67±3.39 $$	39.2±20.1 #
English			
8.00 a.m.	5.19±0.29	10.70±4.95	20.9± 6.8
1.00 p.m.	5.00±0.26	7.40±4.09	38.8±27.5
6.00 p.m.	4.97±0.47	6.41±2.94$	43.2±30.2 $
Milk			
8.00 a.m.	5.25±0.36	8.07±4.00	30.4±15.5
1.00 p.m.	4.94±0.44	4.92±3.65 *	57.2±28.6 *
6.00 p.m.	4.86±0.27 $	6.93±5.35	49.9±29.1 #
All Meals			
8.00 a.m.	5.17±0.34	9.21±4.43	26.3±12.3
1.00 p.m.	4.86±0.39 **	5.66±3.48 **	50.1±27.0 **
6.00 p.m.	4.88±0.41 $$	6.34±3.92 $$	42.0±25.7 $$

Statistical significance of differences: * = 8.00 a.m.–1.00 p.m.; $ = 8.00 a.m.–6.00 p.m.; # = 1.00 p.m.–6.00 p.m., single symbol = $p<0.05$, double symbol = $p<0.01$

found after ingestion at 8.00 a.m. (s. Fig. 1) with highest plasma glucose levels 30 minutes after ingestion (6.17±0.37 mmol/l) and a total duration of approximately 60 minutes. Ingestion at 1.00 p.m. and 6.00 p.m. led to a significantly higher and longer postprandial glycemia. For these meals postprandial plasma glucose reached highest values after 60 minutes (1.00 p.m. meal: 6.57±0.76 mmol/l; 6.00 p.m. meal: 6.88±0.46 mmol/l) and was still significantly elevated 120 minutes after ingestion.

The small glycemic response in the morning was accompanied by markedly enhanced insulin secretion (lower part of fig. 1). After the 8.00 a.m. meal most postprandial insulin was secreted during the first 60 minutes, whereas a more prolonged insulin response was observed after ingestion at 1.00 and 6.00 p.m. Compared to the later meals postprandial plasma insulin levels in the morning were significantly higher from 0 to 45 minutes and significantly lower 90 and 120 minutes after ingestion.

With all test meals the postprandial plasma glucose responses in the afternoon and evening were 10 to 15% higher and approximately twice as long as in the morning (Table 3). Independently of the meal used ingestion at 8.00 a.m. led to about 30% higher plasma insulin levels and maximal values were reached 15% to 20% earlier. Thus no clear influence of meal composition on pattern and magnitude of the diurnal changes could be identified.

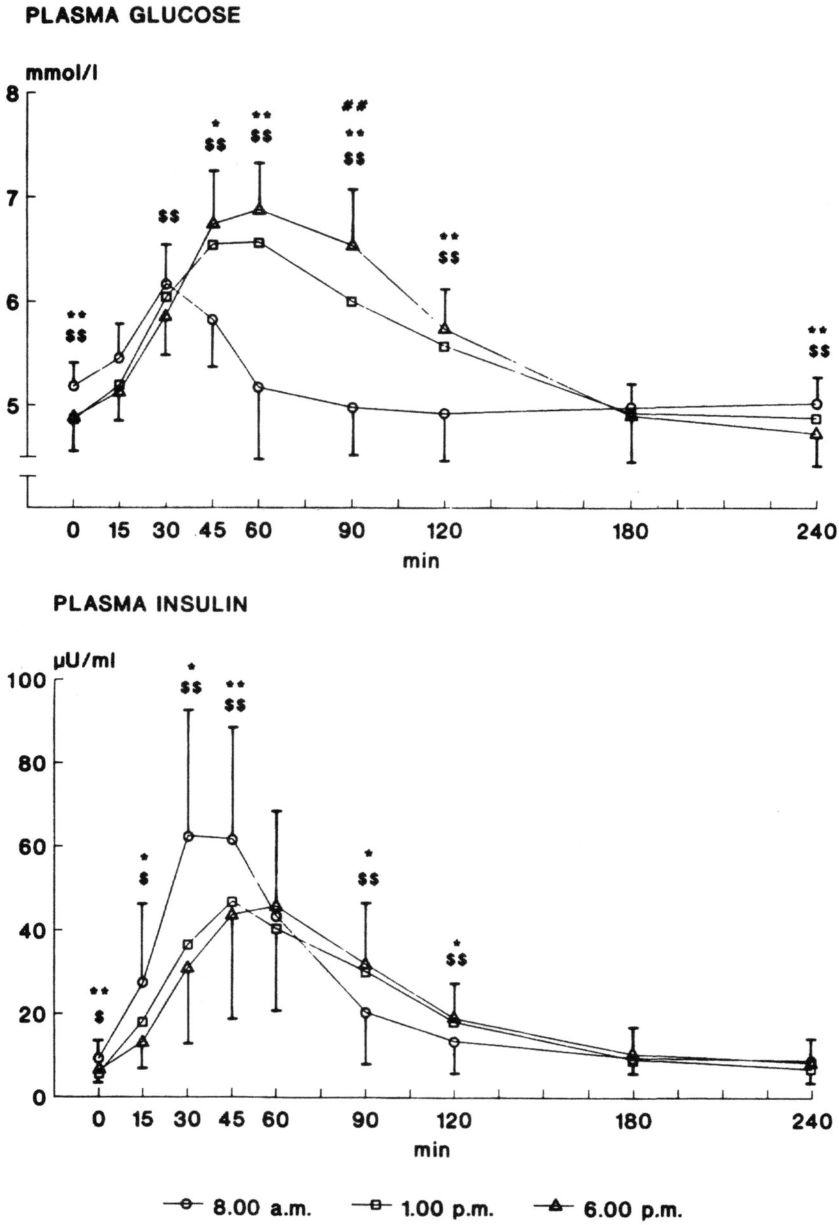

Fig. 1. Postprandial plasma glucose (top) and plasma insulin levels (bottom) in 10 healthy subjects after ingestion at 8.00 a.m., 1.00 p.m., and 6.00 p.m. Means ± SD of all studies (n = 30), without discrimination between the different test meals. Statistical significance of differences: * = 8.00 a.m.–1.00 p.m.; $ = 8.00 a.m.–6.00 p.m.; # = 1.00 p.m.–6.00 p.m., single symbol = $p<0.05$, double symbol = $p<0.01$

Table 3. Maximal values (max.), time of the maximum (t-max.), and total duration of postprandial elevation (t-tot.) of plasma glucose (GL), insulin (I) and insulin sensitivity index (ISI) in 10 healthy subjects after three test meals of equal carbohydrate and energy content ("continental breakfast", "English breakfast", "milk")

	8.00 a.m.	1.00 p.m.	6.00 p.m.
Continental			
max. GL (mmol/l)	6.86 ± 0.83	7.45 ± 0.66 **	7.87 ± 0.68 $$
t-max. GL (min).	35.0 ± 11.5	49.5 ± 14.2 **, #	67.5 ± 16.2 $$
t-tot. GL (min)	95.0 ± 46.4	185.0 ± 46.0 **, #	217.0 ± 14.1 $$
max. I (μU/ml)	75.2 ± 22.9	54.6 ± 23.5 **	62.5 ± 23.0
t-max. I (min).	43.0 ± 8.5	52.5 ± 12.2	66.0 ± 17.6 $$
t-tot. I (min)	150.5 ± 69.5	228.0 ± 40.5 *	228.0 ± 49.4 $$
ISI (l^2/mmol · mU)	2.2 ± 7.6	3.2 ± 2.1	2.4 ± 1.1
English			
max. GL (mmol/l)	6.51 ± 0.66	7.28 ± 1.08 *	7.16 ± 0.75 $$
t-max. GL (min).	34.5 ± 7.2	59.0 ± 15.4 **	67.3 ± 20.3 $$
t-tot. GL (min)	83.0 ± 25.7	156.0 ± 62.9 *	198.0 ± 58.7 $$
max. I (μU/ml)	82.3 ± 65.8	58.3 ± 41.6 #	50.1 ± 25.5
t-max. I (min).	49.5 ± 10.1	60.0 ± 21.2 *	76.5 ± 26.9
t-tot. I (min)	182.0 ± 75.5	216.0 ± 84.6	239.0 ± 67.4 $$
ISI (l^2/mmol · mU)	3.4 ± 2.6	3.4 ± 2.0	3.5 ± 1.6
Milk			
max. GL (mmol/l)	5.89 ± 0.51	6.28 ± 0.52 *, ##	6.84 ± 0.68 $$
t-max. GL (min).	28.5 ± 13.1	39.0 ± 14.5	43.5 ± 19.3 $
t-tot. GL (min)	53.5 ± 26.0	106.0 ± 42.7 *	137.0 ± 35.6 $$
max. I (μU/ml)	82.9 ± 50.1	56.3 ± 29.4	57.2 ± 41.7
t-max. I (min).	31.5 ± 4.7	36.0 ± 10.5	36.0 ± 10.5
t-tot. I (min)	150.0 ± 42.4	162.0 ± 42.9	173.0 ± 62.9
ISI (l^2/mmol · mU)	2.6 ± 1.1	4.3 ± 3.8	3.9 ± 2.6
All Meals			
ISI (l^2/mmol · mU)	2.7 ± 1.7	3.6 ± 2.7	3.3 ± 1.9

min = minutes after ingestion. Statistical significance of differences: * = 8.00 a.m.–1.00 p.m.; $ = 8.00 a.m.–6.00 p.m.; # = 1.00 p.m.–6.00 p.m., single symbol = $p<0.05$, double symbol = $p<0.01$

The areas under the plasma insulin curves for a period of 4 hours after ingestion (Fig. 2) showed no diurnal variation. For all tests performed these areas amounted to 3.17 ± 1.41 mU · ml^{-1} · min for 8.00 a.m., 3.59 ± 1.76 mU · ml^{-1} · min for 1.00 p.m., and 3.55 ± 1.85 mU · ml^{-1} · min for 6.00 p.m. ingestion. No significant differences were found between the different times of ingestion nor between the different test meals.

The insulin sensitivity index (Table 3) did not significantly vary during the day course neither in all meals nor in any paticular.

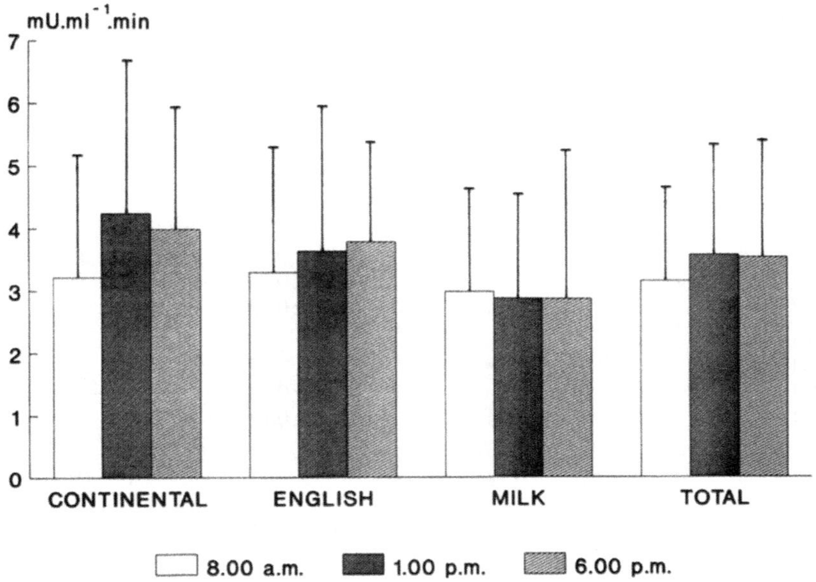

Fig. 2. Areas under the postprandial plasma insulin curves of 10 healthy subjects for a period of 4 hours after ingestion at 8.00 a.m., 1.00 p.m., and 6.00 p.m. Means ±SD for "continental breakfast", "English breakfast", "milk" and pooled for all test meals

2. Type-1 Diabetics. In diabetic subjects preprandial blood glucose and insulin requirements were slightly higher at 8.00 a.m. compared to 1.00 and 6.00 p.m., but differences were small and statistically insignificant (Table 4). Considering both magnitudes by the insulin sensitivity index (Table 4) a difference between lower morning and higher 1.00 p.m. sensitivity was seen. This parallels the findings in healthy subjects (Table 2).

Mean basal insulin requirements during the night and early morning were 2.11±2.21 IU (0.7 IU/h) from 0.00 to 3.00 a.m. and 3.54±2.75 IU (1.18 IU/h) from 5.00 to 8.00 a.m. ($p<0.0001$). An increase in insulin requirements during the early morning was found on 28 out of 30 days studied. Only on three occasions early morning insulin requirements were equal or slightly lower than during the night period. The presence or missing of a dawn phenomenon was not bound to a particular person.

Likewise the diurnal variation of glycemic responses in type-1 diabetic patients was not as pronounced as in non-diabetics, but showed a similar pattern (Fig. 4 top). As in the latter group the 6.00 p.m. meals led to a prolonged postprandial glycemia, and 90 minutes after food ingestion blood glucose was higher (7.82±0.80 mmol/l) than after the 1.00 p.m. (7.24±0.52 mmol/l, $p<0.01$)) and 8.00 a.m. (6.8±1.28 mmol/l, $p<0.05$) meals.

This deterioration of glucose tolerance in the course of the day was confirmed by the insulin infusion rate of the Biostator (Fig. 4, bottom). Between

Table 4. Preprandial blood glucose (BG), insulin requirements (I) and insulin sensitivity index (1000/BG · I) of 10 type-1 diabetic patients for the three test meals ("continental breakfast", "English breakfast", "milk") and for all meals pooled.

Test Meal Time	Blood Glucose mmol/l	Insulin Requirements U/h	Insulin Sensitivity Index (l · h/mmol · U)
Continental			
8.00 a.m.	5.67 ± 0.85	1.05 ± 0.64	0.2 ± 0.2
1.00 p.m.	5.07 ± 1.04	0.75 ± 0.89	1.0 ± 1.3
6.00 p.m.	5.38 ± 0.60	0.66 ± 0.53	0.4 ± 0.3
English			
8.00 a.m.	6.09 ± 0.78	1.86 ± 2.21	0.2 ± 0.4
1.00 p.m.	6.06 ± 1.20	1.33 ± 0.91	0.38 ± 0.46 #
6.00 p.m.	6.00 ± 0.94	1.43 ± 0.81	0.5 ± 0.8 $
Milk			
8.00 a.m.	6.39 ± 0.79	1.35 ± 0.81	0.2 ± 0.2
1.00 p.m.	5.86 ± 0.86	1.58 ± 1.09	0.7 ± 1.4
6.00 p.m.	5.99 ± 1.00	1.60 ± 1.05	0.7 ± 1.3
All Meals			
8.00 a.m.	6.03 ± 0.85	1.42 ± 1.40	0.2 ± 0.2
1.00 p.m.	5.67 ± 1.10	1.23 ± 0.99	0.7 ± 1.1 *
6.00 p.m.	5.79 ± 0.90	1.23 ± 0.90	0.5 ± 0.9

Statistical significance of differences: * = 8.00 a.m.–1.00 p.m.; $ = 8.00 a.m.–6.00 p.m.; # = 1.00 p.m.–6.00 p.m., single symbol = $p<0.05$, double symbol = $p<0.01$

45 and 120 minutes after ingestion the 6.00 p.m. meals required distinctly higher insulin infusion rates to maintain normoglycemia.

This pattern of circadian variation could be seen with all three test meals, but unlike the situation in healthy subjects the magnitude of the diurnal changes was clearly affected by the composition of the test meal (Table 5). For most parameters differences between ingestion times were largest for "continental breakfast" or "milk" and smallest for the "English breakfast", which had the highest 8.00 a.m. insulin sensitivity index (3.4 ± 2.6 l^2/mmol · mU) in healthy subjects at breakfast time and the highest dietary fiber content. The influence of the test meal was most evident on the total duration of postprandial elevation of the insulin infusion rate (t-tot. IR). The increase in this parameter between ingestion at 8.00 a.m. and at 6.00 p.m. amounted to 75% with "milk", 43% with "continental breakfast" and 24% for "English breakfast".

Mean postprandial (4 h) insulin requirements significantly increased in the course of the day. For all tests performed mean insulin requirements amounted to 8.49 ± 2.98 IU for the 8.00 a.m meal, 9.1 ± 1.98 IU for the 1.00 p.m. meal and 10.87 ± 2.97 IU for the 6 p.m. meal (Fig. 4). Again the diurnal changes were smallest and statistically not significant for the "English breakfast".

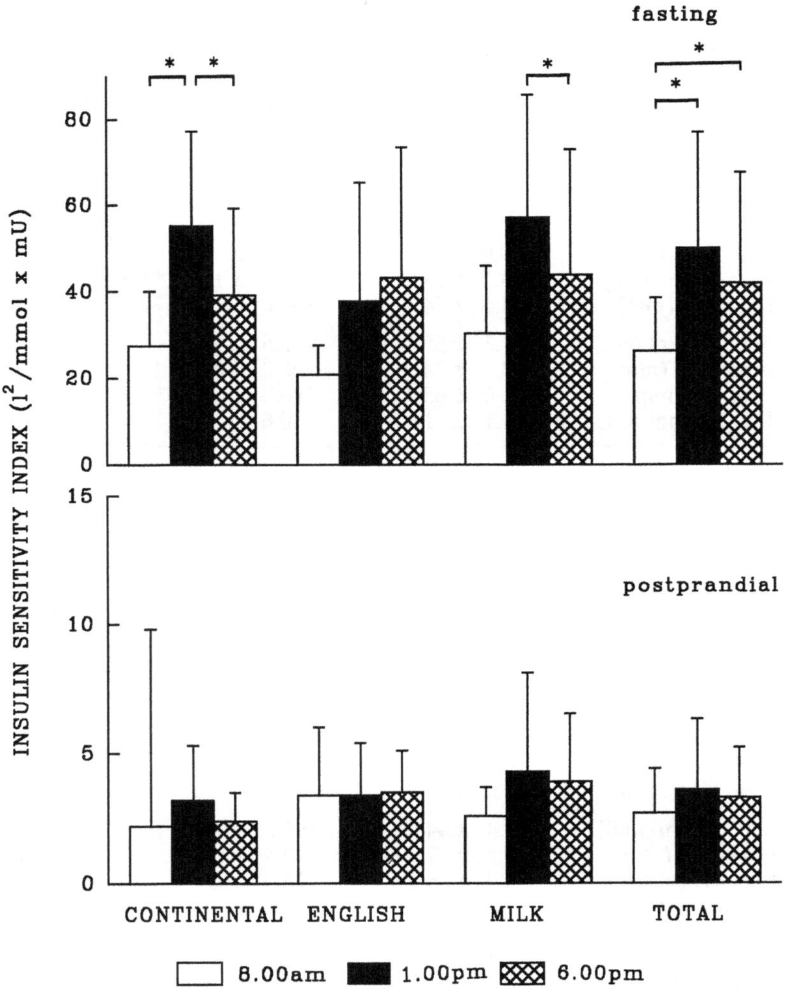

Fig. 3. Insulin sensitivity index (ISI) of 10 healthy subjects. Mean ±SD for "continental breakfast", "English breakfast", "milk" and pooled for all test meals.
1. Preprandial ISI = 1000/BG · I (l^2/mmol · mU); BG = blood glucose, I = insulin level
2. Postprandial ISI = $1000/BG_{max}/I_{max}$ (l^2/mmol · mU); BG_{max} = maximal blood glucose, I_{max} = maximal insulin level after food ingestion

Discussion

Our results in normal individuals confirm for all test meals a pronounced deterioration of glucose tolerance in the afternoon and evening [1–7, 10, 11]. The better glucose tolerance in the morning is believed to be due to enhanced insulin secretion [2, 3, 5–9] and higher peripheral insulin sensitivity [5, 7, 10, 11]. Enhanced insulin secretion in the morning was clearly detectable in the

Table 5. Maximal values (max.), time of the maximum (t-max.), and total duration of postprandial elevation (t-tot.) of plasma glucose (GL), insulin infusion rate (IR) and insulin sensitivity index (1000/GL · IR) in 10 type-1 diabetic patients after three test meals of equal carbohydrate and energy content ("continental breakfast", "English breakfast", "milk")

	8.00 a.m.	1.00 p.m.	6.00 p.m.
Continental			
max. GL (mmol/l)	9.06 ± 1.81	9.12 ± 1.35	9.60 ± 1.29
t-max. GL (min)	64.5 ± 34.0	60.0 ± 17.8 **, ##	67.5 ± 16.2
t-tot. GL (min)	144.5 ± 54,4	183.0 ± 43.5	213.0 ± 45,7 $$
max. IR (mU/ml)	177.3 ± 55.7	202.5 ± 53.2	223.1 ± 84.4
t-max. IR (min).	43.5 ± 24.9	60.0 ± 23.5 *	57.0 ± 19.7
t-tot. IR (min)	155.0 ± 85.4	196.0 ± 61.9 *	222.0 ± 47.3 $$
ISI (h · l/mmol · U)	1.1 ± 1.0	0.6 ± 0.5	0.5 ± 0.3
English			
max. GL (mmol/l)	8.33 ± 0.82	8.43 ± 0.96	8.51 ± 0.64
t-max. GL (min)	49.5 ± 10.1	63.0 ± 21.0	64.5 ± 18.8 $$
t-tot. GL (min)	114.0 ± 38.9	140.0 ± 41.8	167.0 ± 68.7
max. IR (mU/ml)	130.0 ± 41.7	134.4 ± 74.9	141.0 ± 54.4
t-max. IR (min)	42.0 ± 9.50	45.0 ± 12.2	43.5 ± 11.1
t-tot. IR (min)	125.0 ± 39.8	165.5 ± 61.2	155.0 ± 57.0
ISI (h · l/mmol · U)	1.0 ± 0.6	1.2 ± 0.9	1.0 ± 0.6
Milk			
max. GL (mmol/l)	8.97 ± 0.61	9.04 ± 0.85	9.46 ± 0.76
t-max. GL (min)	44.0 ± 5.20	49.5 ± 17.4	51.0 ± 7.8
t-tot. GL (min)	120.0 ± 67.8	159.0 ± 44.8	163.0 ± 49.9
max. IR (mU/ml)	184.4 ± 44.6	196.6 ± 65.5	198.2 ± 34.1
t-max. IR (min)	37.5 ± 7.9	40.5 ± 18.8	40.5 ± 7.2
t-tot. IR (min)	94.0 ± 46.2	125.0 ± 59.3	165.0 ± 43.0 $$
ISI (h · l/mmol · U)	0.7 ± 0.3	0.7 ± 0.4	0.6 ± 1.1
All Meals			
ISI (h · l/mmol · U)	0.9 ± 0.7	0.8 ± 0.6	0.7 ± 0.4

min = minutes after ingestion. Statistical significance of differences: * = 8.00 a.m.–1.00 p.m.; $ = 8.00 a.m.–6.00 p.m.; # = 1.00 p.m.–6.00 p.m., single symbol = $p<0.05$, double symbol = $p<0.01$

present study. Ingestion at 8.00 a.m. led to highest plasma insulin levels and to a significantly faster insulin secretion than in the afternoon and evening. The rapid insulin release in the morning could be caused by enhanced sensitivity of β-cells to secretory stimuli [38] and/or higher pancreatic content of quickly releasable insulin after overnight fast [8].

No circadian variation was found between the areas under the postprandial plasma insulin curves for a period of 4 hours. This finding indicates that the total amount of postprandial insulin did not differ significantly at different times of the day. The fact that equal amounts of insulin induce significant-

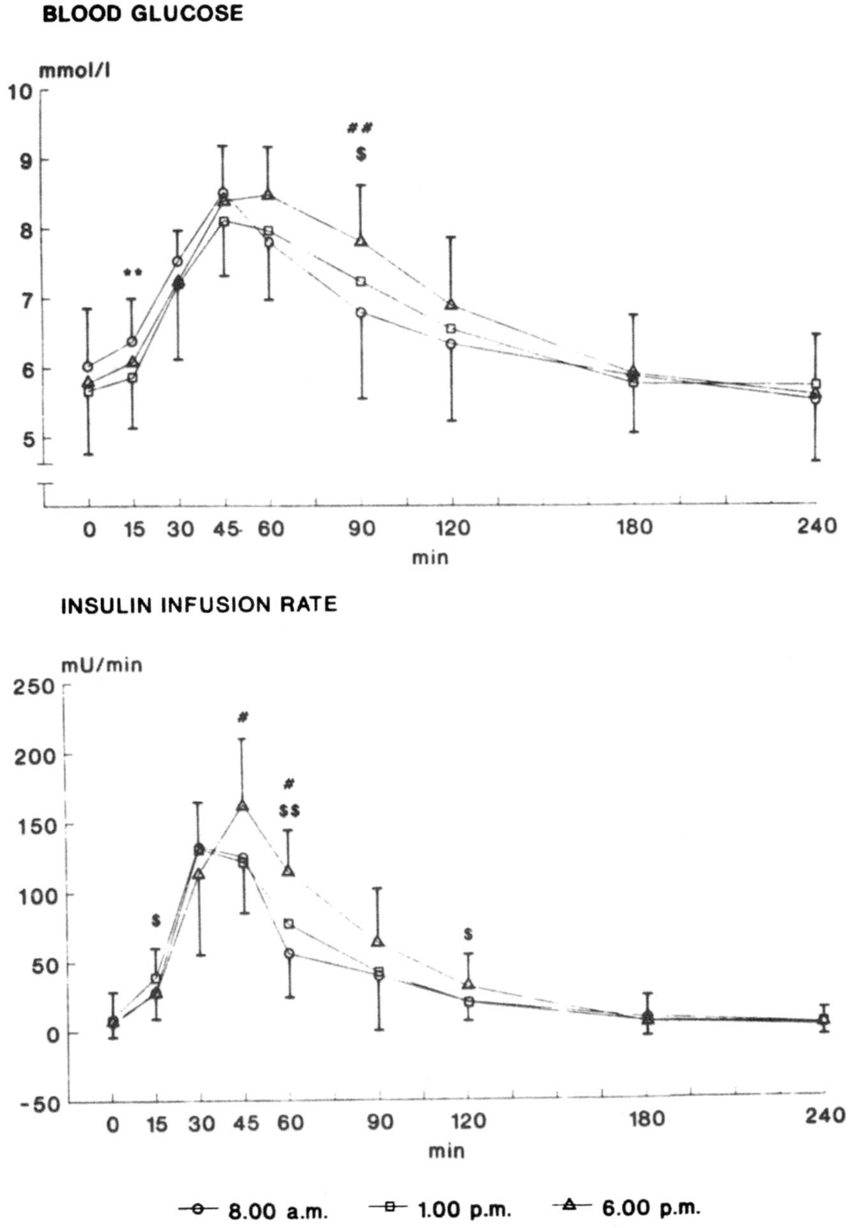

Fig. 4. Postprandial blood glucose (top) and insulin infusion rates (bottom) in 10 type-1 diabetic patients after ingestion at 8.00 a.m., 1.00 p.m., and 6.00 p.m. Means ± SD of all studies (n = 30), without discrimination between the different test meals. Statistical significance of differences: * = 8.00 a.m.–1.00 p.m.; $ = 8.00 a.m.–6.00 p.m.; # = 1.00 p.m.–6.00 p.m., single symbol = $p<0.05$, double symbol = $p<0.01$

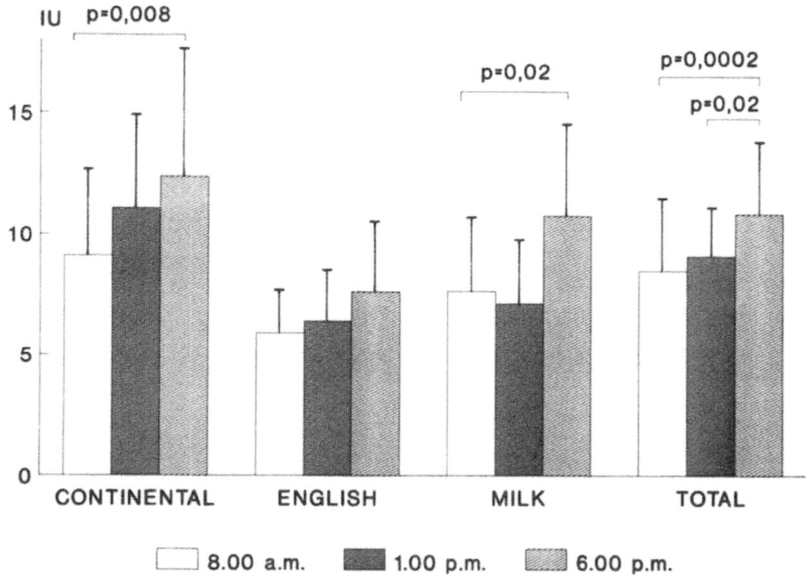

Fig. 5. Postprandial insulin requirements of 10 type-1 diabetic patients for a period of 4 hours after ingestion at 8.00 a.m., 1.00 p.m., and 6.00 p.m. Means ±SD for "continental breakfast", "English breakfast", "milk" and pooled for all test meals

ly smaller glycemic responses in the morning supports findings of a progressive decrease in insulin sensitivity in the course of the day, although the insulin sensitivity index used showed no significant differences (Table 2). This relative insulin resistance in the evening may be due to higher levels of free fatty acids [3, 6], diurnal changes in growth hormone levels [4, 39–41], and possibly a down-regulation of insulin receptors [42].

The same amount of insulin is more effective in form of an intravenous bolus than of a slow, continuous infusion [43, 44]. In addition to enhanced insulin sensitivity the rapid insulin secretion after breakfast is therefore likely to contribute directly to the higher hypoglycemic power of the morning insulin.

The existence of an increase in basal glucose and insulin levels during the early morning is still controversial concerning healthy man. In correspondence to the dawn phenomenon of diabetic patients some studies found increased pancreatic insulin secretion, relative insulin resistance and increased hepatic gluconeogenesis of normal humans in the early morning [45, 46]. Supporting these findings healthy subjects in the present study had 5% higher preprandial plasma glucose and 30% higher plasma insulin at 8.00 a.m. compared to 1.00 p.m. and 6.00 p.m. This relative insulin resistance in the morning would antagonize the observed diurnal changes of postprandial glycemic response, indicating that circadian variations of basal and postprandial glucose tolerance must be unrelated and due to different mechanisms.

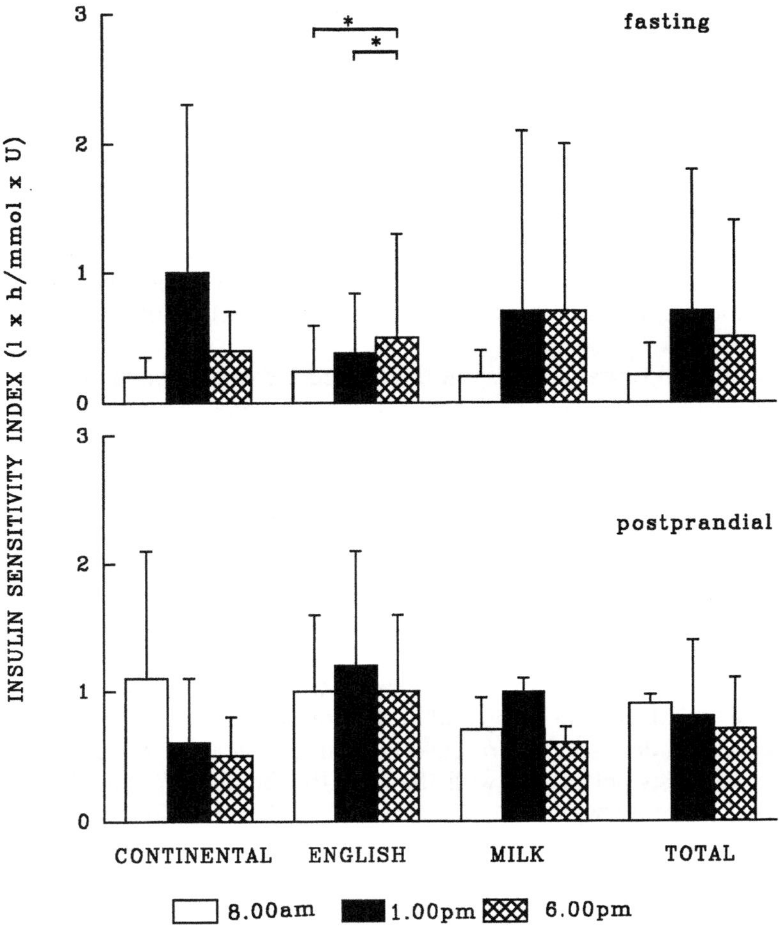

Fig. 6. Insulin sensitivity index (ISI) of 10 type-1 diabetic patients. Mean ±SD for "continental breakfast", "English breakfast", "milk" and pooled for all test meals.

1. Preprandial ISI = 1000/BG · IR (l · h/mmol · U); BG = blood glucose, IR = insulin infusion rate at 8.00 a.m., 1.00 p.m. and 6.00 p.m.
2. Postprandial ISI = $1000/BG_{max} \cdot IR_{max}$ (l · h/mmol · U); BG_{max} = maximal blood glucose, IR_{max} = maximal insulin infusion rate after food ingestion at 8.00 a.m., 1.00 p.m. and 6.00 p.m.

The postprandial results in type-1 diabetic patients clearly contradict previous findings indicating that in these patients highest insulin requirements per gram carbohydrate are found for breakfast [12–17]. In the present study postprandial blood glucose responses and insulin requirements of the diabetic patients showed a progressive increase in the course of the day. Compared to ingestion at 8.00 a.m. mean postprandial insulin requirements for identical test meals were 7% higher at 1.00 p.m. and 28% higher at 6.00 p.m. Thus under the conditions of the present investigation the type-1 diabetic patients showed a circadian behavior similar to non-diabetic patients with a reduction of glucose tolerance during the day.

Some previous studies performed in insulin dependent diabetic patients with intravenous insulin delivery systems also failed to confirm highest insulin requirements per gram carbohydrate for breakfast [3, 26–30]. Subcutaneous insulin inevitably leads to a prolonged peripheral hyperinsulinemia. This excess of circulating insulin is likely to reduce the insulin requirements of the following meal. The intravenous insulin delivery in combination with the small size of the test meals and the large intervals between ingestion in the present study may have been crucial for detection of the observed diurnal changes.

The postprandial response to mixed meals with equal amounts of carbohydrates may vary considerably [47, 48, 32], depending on factors like the nature of the carbohydrate [49] or the content of fat [50, 51] and dietary fiber [52, 53]. Independently of the time of the day the "English breakfast" of the present study required between 35% and 45% less insulin than the "continental breakfast" (Fig.4) although both meals had an identical carbohydrate and energy content. The fact that food usually eaten for breakfast has a higher glycemic effect and higher insulin requirements than the meals eaten for lunch and dinner [32, 48] will contribute to higher morning insulin requirements under daily life conditions.

Postprandial insulin requirements are also known to be influenced by the size [29] and composition [54] of the preceding meal. In the present study the circadian deterioration of postprandial carbohydrate tolerance was smallest in the tests performed with the "English breakfast". This finding supports previous observations that meals with low glycemic effect and insulin requirements have a favourable effect on the insulin requirements of the following meal [54].

As in healthy subjects the circadian variation of basal glucose metabolism of diabetic patients was independent of the observed circadian pattern of postprandial responses. Basal insulin requirements were lowest during the night, increased considerably during the early morning and remained fairly stable throughout daytime. Though in the present study a dawn phenomenon [18, 21, 23, 24, 55, 56] was almost always present it evidently did not lead to a reduction of postprandial carbohydrate tolerance in the morning. Accordingly the deterioration of postprandial carbohydrate tolerance in the evening was not accompanied by an increase in fasting insulin requirements.

We conclude that in healthy men as in type-l diabetic patients the diurnal variation of postprandial carbohydrate tolerance and insulin sensitivity, respectively are independent from the rhythmics of basal glucose metabolism. In diabetic patients the circadian pattern of postprandial responses is substantially influenced by the way of insulin delivery, caloric intake and the composition of the diet and the time interval between meals. In subcutaneously treated diabetic patients eating a normocaloric mixed diet these external factors are able to mask completely the diurnal deterioration of postprandial carbohydrate tolerance demonstrated in the present study.

Acknowledgements. We thank Professor N-G. Asp from the Department of Food Chemistry of the University of Lund, Sweden, for the analysis of the carbohydrate and fiber content of the test meals. This study was supported by the award of a H. & L. Schilling Professorship by the "Stifterverband für die Deutsche Wissenschaft" to the first author of this article.

References

1. Bowen AJ, Reeves RL (1967) Diurnal variation in glucose tolerance. Arch Intern Med 119:261–264.
2. Jarrett J, Baker IA, Keen H, Oakley NW (1972) Diurnal variation in oral glucose tolerance: blood sugar and plasma insulin levels morning, afternoon, and evening. Br Med J 22:199–201.
3. Carroll KF, Nestel PJ (1973) Diurnal variation in glucose tolerance and in insulin secretion in man. Diabetes 22:333–348.
4. Aparicio NJ, Felix E, Puchulu E, Gagliardino JJ, Ruiz M, Lorens JM, Ruiz J (1974) Circadian variation of the blood glucose, plasma insulin and human growth hormone levels in response tó an oral glucose load in normal subjects. Diabetes 23:132–137.
5. Whichelow J, Sturge RA, Keen H, Jarrett RJ, Stimmler L, Grainger S (1974) Diurnal variation in response to intravenous glucose, Br Med J 1:488–491.
6. Zimmet PZ, Wall JR, Rome R, Stimmler L, Jarrett RJ (1974) Diurnal variation in glucose tolerance: associated changes in plasma insulin, growth hormone and non-esterified fatty acids. Br Med J 1:485–491.
7. Tasaka Y, Sekine M, Wakatsuki M, Ohgawara H, Shizume K (1975) Levels of pancreatic glucagon, insulin and glucose during twenty-four hours of the day in normal subjects. Horm Metab Res 7:205–206.
8. Malherbe C, Gasparo des M, Herthogh des R, Hoet JJ (1969) Circadian variations of blood sugar and plasma insulin levels in man. Diabetologia 5:397–404.
9. Polonsky KS, Given BD, Cauter van E (1988) Twenty-four-hour profiles and pulsatile patterns of insulin secretion in normal and obese subjects. J Clin Invest 81: 442–448.
10. Nemeth S, Vigas M, Macho L, Stukowsky R (1970) Diurnal variation of the disappearence rate of glucose from the blood of healthy subjects as revealed by intravenous glucose tolerance tests. Diabetologia 6:641.
11. Gibbson T, Jarrett RJ (1972) Diurnal variation in insulin sensitivity. Lancet I: 947–948.
12. Hepp KD, Renner R, Funcke von HJ, Mehnert H (1977) Glucose homoeostasis under continuous intravenous insulin therapy in diabetics. Horm Metab 7:72–76 (Suppl.).
13. Tamborlane WV, Sherwin RS, Genel M, Felig P (1979) Reduction to normal of plasma glucose in juvenile diabetes by subcutaneous administration of insulin with a portable infusion pump. N Engl J Med 300:573–578.
14. Rizza RA, Gerich JE, Haymond MW, Westland RE, Hall LD, Clemens AH, Service FJ (1980) Control of blood sugar in insulin-dependent diabetes: comparison of an artificial endocrine pancreas, continuous subcutaneous insulin infusion, and intensified conventional insulin therapy. N Engl J Med 303:1313–1318.
15. Kerner W, Schultz M, Schock D, Pfeiffer EF (1982).Variations of insulin requirements in insulin dependent diabetics for meals taken at different times of the day. Horm Metab Res, Suppl. 12.
16. Mathiesen ER, Rubin P, Sandahl Christiansen J, Aaby Svendsen P, Lauritzen T, Deckert T (1982) Diurnal pattern of insulin requirements in insulin-dependent diabetics. Scand J Cin Lab Invest 42:63–68.

17. Raskin P (1982) Treatment of insulin-dependent diabetes mellitus with portable insulin delivery devices. Med Clin N Am 66:1269–1283.
18. Schrezenmeir J, Tato F, Tato S, Achterberg H, Schulz G, Strack T, v. Aerssen M, Kasper H, Beyer J (1985) Food-dependent insulin consumption. In: Beyer J, Albisser M, Schrezenmeir J, Lehmann L (eds.) Computer systems for insulin adjustment in diabetes mellitus. Panscientia, Hedingen, pp 43–53.
19. Skyler JS, Seigler DES, Reeves ML (1982) Optimizing pumped insulin delivery. Diabetes Care 5:135–147.
20. Schrezenmeir J, Tato F, Tato S, Müller-Haberstock S, Achterberg H, Stürmer W, Hogan M, Beyer J (1987) Algorithms for CSII Quantified approach to basal rate and meal-induced requirements. In: Brunetti P, Waldhäusl WK, (eds.) Advanced models for the therapy of insulin-dependent diabetes. Serono Symposia Publications from Raven Press, New York, vol. 37:191–197
21. Clarke WL, Haymond MW, Santiago JV (1980) Overnight basal insulin requirements in fasting insulin-dependent diabetics, Diabetes 29:78–80.
22. Campbell PJ, Bolli GB, Cryer PE, Gerich JE (1984) Nocturnal spikes in growth hormone secretion cause the dawn phenomenon. Diabetologia 27:262A.
23. Campbell PJ, Bolli GB, Cryer PE, Gerich JE (1985) Pathogenesis of the dawn phenomenon in patients with insulin-dependent diabetes mellitus. N Engl J Med 312:1473–1479.
24. Kerner W, Navascues I, Torres AA, Pfeiffer EF (1984) Studies on the pathogenesis of the dawn phenomenon in insulin-dependent diabetic patients. Metabolism 33:458–464.
25. Koivisto VA, Yki-Yarvinen H, Helve E, Karonen S-L, Pelkonen R (1986) Pathogenesis and prevention of the dawn phenomenon in diabetic patients treated with CSII. Diabetes 35:78–82.
26. Perlman K, Ehrlich RM, Filler RM, Albisser M (1981) Waveform requirements for metabolic normalization with continuous intravenous insulin delivery in man. Diabetes 30:710–717.
27. Slama G, Klein J-C, Delage A, Ardila E, Lemaignen H, Papoz L, Tchobroutsky G (1981) Correlation between the nature and amount of carbohydrate in meal intake and insulin delivery by the artificial pancreas in 24 insulin-dependent diabetics. Diabetes 30:101–105.
28. Schulz G, Beyer J, Happ J, Cordes U, Küstner E, Hassinger W (1982) Variation of basal and carbohydrate dependent insulin consumption during a 24 hour period of control. Horm Metab Res 12:221–224 (Suppl).
29. Service J, Rizza RA, Hall LD, Westland RE, O'Brien PC, Clemens H, Haymond MW, Gerich JE (1983) Prandial insulin requirements in insulin-dependent diabetics: effects of size, time of day, and sequence of meals. J Clin Endocrinol Metab 57 :931–936.
30. Rewers M, Dmochowski K, Walczak M (1985) Diurnal variation in carbohydrate tolerance to mixed meal in insulin-dependent diabetic adolescents during continuous intravenous insulin infusion, Endokrinologia Polska 36:1 P1 ISSN 0423–104X.
31. Souci SW, Fachmann W, Kraut H (1981) Die Zusammensetzung der Lebensmittel, Nährwerttabellen 1981/82. Wissenschaftliche Verlagsgesellschaft Stuttgart.
32. Schrezenmeir J, Tato F, Tato S, Küstner E, Krause U, Hommel G, Asp N-G, Kasper H, Beyer J (1989) Comparison of glycemic response and insulin requirements after mixed meals of equal carbohydrate content in healthy, type-l, and type-2 diabetic man. Klin Wochenschr 67:985–994.
33. Holm J, Bjork I, Drews A, Asp N-G (1986) A rapid method for the analysis of starch. Starch 28:224–226.
34. Asp N-G, Johansson C-G, Hallmer H, Siljestrom M (1983) Rapid enzymatic assay of insoluble and soluble dietary fiber, J Agric Food Chem 31:476–482.
35. Clemens AH, Chag PH, Myers RW (1977) The development of Biostator, a glucose controlled insulin infusion system (GCIIS). Horm Metab Res (Suppl Ser 7).

36. Hales CN, Randle PJ (1963) Immunoasaay of insulin with an insulin-antibody precipitate. Biochem J 88:137–146.
37. Desbuquois B, Aurbach GD (1971) Use of polyethylene glycol to separate free and antibody-bound peptide hormones in radioimmunoassays. Clin Endocrinol 33:732–738.
38. Gagliardino JJ, Hernandez RE, Gagliardino EE (1972) Relationship between differential responsiveness of pancreatic β-cells and the circadian variation of serum immunoreactive insulin levels. Endocrinology 91:822–825.
39. Quabbe HJ, Schilling E, Helge H (1966) Pattern of growth hormone secretion during a 24-hour fast in normal adults. J Clin Endocrinol 26:1173–1177.
40. Laan van der WP, Parker DC, Rossman LG, Laan van der EF (1970) Implications of growth hormone release in sleep. Metabolism 19:891–897.
41. Bratusch-Marrain PR, Smith D, Fronzo des RA (1982) The effect of growth hormone on glucose metabolism and insulintion in man. J Clin Endocrinol Metab 55:973–982.
42. Beck-Nielsen H, Pederson O (1978) Diurnal variation in insulin binding to human monocytes. J Clin Endocrinol Metab 47:385–390.
43. Strack T, Krause U, Schulz G, Beyer J, Beutelsbacher F, Nagel J (1984) Development of individual insulin infusion profiles for open loop infusion systems. Horm Metab Res 16:163–167.
44. Matthews DR, Naylor BA, Jones RG, Ward GM, Turner RC (1983) Pulsatile insulin has greater hypoglycemic effect than continuous delivery. Diabetes 32:617–621
45. Bolli GB, Des Feo P, Des Cosmo S, Perriello G, Ventura MM, Calcinaro F, Lolli C, Campbell P, Brunetti P, Gerich JE (1984) Demonstration of a dawn phenomenon in normal human volunteers. Diabetes 33:1150–1153
46. Schmidt MI, Xiong Lin Q, Gwynne JT, Jacobs S (1984) Fasting early morning rise in peripheral insulin: evidence of the dawn phenomenon in nondiabetes. Diabetes Care 7:32–35.
47. Schauberger G, Brinck VC, Guldner G, Spaethe R, Niklas L, Otto H (1978) Exchange of carbohydrates according to their effect on blood glucose. Diabetes 26: 415–420.
48. Jenkins DJA, Wolever TMS, Taylor RH, Barker H, Fielden H, Baldwin JM, Bowling AC, Newman HC, Jenkins AL, Goff DV (1981) Glycemic index of foods: a physiological basis of carbohydrate exchange. Am J Clin Nutr 34:262–366.
49. Crapo PA, Reaven G, Olefsky J (1976) Plasma glucose and insulin responses to orally administered simple and complex carbohydrates. Diabetes 25:741–747.
50. Estrich D, Ravnik A, Schlierf G, Fukayama G, Kinsell L (1967) Effects of coingestion of fat and protein upon carbohydrate induced hyperglycemia. Diabetes 16: 232–237.
51. Wakhloo AK, Beyer J, Diederich C, Schulz G (1984) Einfluß von Nahrungsfett auf Blutzuckerspiegel und Insulinverbrauch nach Einnahme verschiedener Kohlenhydratträger bei TypI-Diabetikern am künstlichen Pankreas. Dtsch Med Wschr 109:1589–1594.
52. Jenkins DJA, Leeds AR, Wolever TMS, Goff DV, Alberti KGMM, Gassull MA, Hockaday TD (1976) Unabsorbable carbohydrates and diabetes: decreased postprandial hyperglycemia. Lancet I:172–174.
53. Goulder TJ, Alberti KGMM, Jenkins DJA (1978) Effect of added fibre on the glucose and metabolic response to a mixed meal in normal and diabetic subjects. Diabetes Care 1:351–355.
54. Schmidt MI, Hadji-Georgopoulos A, Rendell M, Margolis S, Kowarski A (1981) The dawn phenomenon, an early morning glucose rise: implications for diabetic intraday blood glucose variation. Diabetes Care 4:579–585.
55. Bolli GB, Gerich JE (1984) The "dawn phenomenon" – a common occurrence in both non-insulin-dependent and insulin-dependent diabetes mellitus. N Engl J Med 310:746–750.

56. Nestler JE, Barlascini CO, Clore JN, Blackard WG (1988) Absorption characteristics of breakfast determines insulin sensitivity and carbohydrate tolerance for lunch. Diabetes Care 11:755–756.

Effect of Fish Oil Concentrate on the Lipoprotein Profile in Patients Suffering from Diabetes mellitus Type II

W. HERRMANN, J. BIERMANN, K.-P. RATZMANN, and H.-G. LINDHOFER

Unfavourable changes of the lipoprotein metabolism are reported in patients suffering from diabetes mellitus type II, which are partly responsible for the increased atherogenic risk in these patients [1, 6]. The endogenous synthesis of the very-low-density (VLDL)-triglycerides [13] and of the apolipoprotein B (apo B) [4] is increased and results in a pathological serum level of triglycerides. Low-density-lipoprotein (LDL)-apo B is also raised and linked with an unfavourable LDL-composition [5]. The overproduction of VLDL and LDL-apo B leads to premature coronary artery disease even in patients without hyperlipemia [6].

n-3 fatty acid rich fish oil diet results in a striking decrease of the triglyceride level in normolipemic and hypertriglyceridemic nondiabetics. This effect is probably reached by inhibition of the VLDL-triglyceride- and apo B-synthesis [3, 10]. In the case of a high dosage of fish oil, the LDL-apo B-turnover is reduced and the apo B-concentration in serum can also be reduced [8].

Beside the positive effects on the lipoprotein metabolism, n-3 fatty acid rich fish oil diet influences further atherogenic risk factors favourably. The eicosanoid balance is shifted towards the antiatherogenic and vasodilatatoric side [3]. A reduction of the thrombocyte aggregation rate, a decrease of the blood pressure and an improvement of the blood or plasma viscosity are associated with it [3].

As diabetes mellitus type II is associated with increased serum levels of VLDL-triglycerides and VLDL-apo B, the intake of fish oil could be useful in lowering circulating VLDL in patients with diabetes mellitus type II. Therefore we studied the effect of n-3 fatty acid rich diet during 12 weeks of treatment in diabetics of type II who until now have not been treated by drugs but only by diet.

Patients and Methods

The study included 19 male patients suffering from diabetes mellitus type II (NIDDM) (age 47.8 ± 9.4 years). The metabolic disorder had been known for about 4.1 years. All patients were treated by diet only and by medical advice. The metabolism was regularly controlled.

The investigations were carried out as an open study starting with a placebo-run-in-period for four weeks (phase I) (6 g rape seed oil capsules/d), following a twelve week verum period (phase II) (6 g fish oil concentrate capsules/d, FOURNIER PHARNMA GmbH), and concluding with a wash-out-period of four weeks (phase III) (6 g rape seed oil capsules/d).

The lipids, apolipoproteins, glucose, glucose tolerance, and insulin were checked at the beginning and at the end of each phase. The blood glucose was investigated in capillary blood by means of glucose sensitive electrode. The plasma insulin level (IRI) was determined by radioimmunological assay (Isocomerz), the apolipoproteins (apo) A-I and B by immunoturbidimetric test kit (SIFIN Berlin), and the enzyme activities (ALAT, GGT) by kinetic test kit combination (Boehringer). HDL-cholesterol and its subfractions as also LDL-cholesterol were quantified by precipitation techniques (Quantolip® A and B, Quantolip®-LDL, IMMUNO AG).

The patients were instructed to continue their life style recommended by their doctors. The patients practised their professions during the study. The body weight was constant during all test phases.

The statistical calculation was done by significance tests with the t-test by Student, after checking the variation with the F-test or the examination of normal distribution.

Results

The cholesterol level was relatively constant during all test phases. LDL-cholesterol increased during the verum phase by about 10%, but only the difference at the start was significant. In the wash-out-phase, LDL-cholesterol returned to the start level. Apo B increased slightly by about 5%. Only the difference at the beginning was statistically significant. During the wash-out period, apo B remained on the same level as before (Table 1).

HDL-cholesterol increased significantly by about 9%, especially HDL_2-cholesterol. During the wash-out-period, HDL cholesterol showed no further change.

The fish oil treatment markedly lowered apo A-I by about 9%. During the wash-out-phase, apo A-I remained at the same low level.

Fish oil also caused a distinct fall in triglyceride concentration in serum by 29% followed by a strong increase during the wash-out-phase. The fasting blood glucose and the glucose tolerance test as well as the insulin level (fasting and after load test) did not change significantly at the end of the verum phase in comparison to the run-in-period.

Discussion

Our results support the hypothesis that fish oil inhibits the VLDL-and triglyceride-synthesis in the liver. The data profile is similar to that in normolipemics and hyperlipidemic patients after fish oil diet [3, 8–10]. The decrease in

Table 1. Lipoprotein parameters, blood glucose, glucose tolerance, and insulin (fasting and after load test) during supplementary fish oil diet in patients with diabetes mellitus type II.

Lipids & Apolipoporoteins	Start	Phase 1	Phase 2	Phase 3
TC (mmol/l)	7,02±1.76	7.13±1.68	7.25±1.65	7.24±2.65
HDL-C (mmol/l)	1.18±0.24	1.18±0.28	1.29±0.30**/##	1.26±0.36
HDL_3-C (mmol/l)	0.89±0.18	0.91±0.21	0.93±0.18	0.97±0.30**
HDL_2-C (mmol/l)	0.29±0.17	0.27±0.18	0.36±0.20	0.29±0.13
LDL-C (mmol/l)	4.42±1.11	4.64±1.32	5.08±1.45*	4.18±1.28++
TG (mmol/l)	2.73±1.87	2.74±2.19	1.94±1.53*/#	3.69±3.10+
Apo A-I (g/l)	1.54±0.16	1.58±0.17	1.44±0.14**/###	1.40±0.20**/###
Apo B (g/l)	1.44±0.34	1.47±0.35	1.54±0.30*	1.53±0.30*
LDL-C/HDL-C	3.99±1.24	4.21±1.66***	4.14±1.56***	3.34±1.21***/###
Glucose-Tolerance				
Blood glucose (mmol/l)				
– Beginning	6.5± 1.7	7.4± 2.4*	7.5± 2.5	7.6± 2.6
– 30 minutes	11.3± 2.7	13.5± 3.6***	12.7± 3.1*	12.6± 4.1
– 60 minutes	14.0± 3.9	17.1± 4.2***	16.3± 4.7**	16.2± 4.8*
– 120 minutes	11.1± 4.5	12.5± 4.7	11.7± 5.4	14.4± 5.9*/++
Insulin				
(mU/l)				
– Beginning	10.1± 6.5	9.5± 5.2	11.7± 7.1	9.2± 5.3
– 30 minutes	32.8±26.9	26.1±20.1	33.7±24.3	35.7±27.7#
– 60 minutes	55.2±42.0	54.4±50.3	56.0±48.6	53.4±33.7
– 120 minutes	40.2±35.5	34.8±27.1	37.6±19.3	39.0±22.1

Phase 1: Placebo-Run-In-Phase; during 4 weeks; *Phase 2:* Verum-Phase; during 12 weeks; *Phase 3:* Washout-Phase; during 4 weeks; * = significance to start; # = significance to phase 1; + = significance to phase 2; */#/+ = probability level 0.05; **/##/++ = probability level 0.01; ***/###/+++ = probability level 0.001
TC = Total-cholesterol; HDL-C = HDL-cholesterol; LDL-C = LDL-cholesterol; TG = Triglycerides; Apo A-I = Apolipoprotein A-I; Apo B = Apolipoprotein B

VLDL-cholesterol and VLDL-triglycerides which can be seen during n-3 PUFA-rich diet bases on the diminution of circulating VLDL.

In spite of this VLDL-and triglyceride-lowering effect our study confirms like other studies [4–5, 11–12] a slight increase in LDL-cholesterol and apo B at the end of the fish oil period. This effect seems to be paradoxical because fish oil inhibits the synthesis of VLDL-triglycerides and -apo B in the liver and as the main metabolic pathway of VLDL is the conversion to LDL.

The different findings from the literature make clear that the effect on LDL-cholesterol and apo B depends on the fish oil dosage and on the underlying lipid metabolic disorder [3, 6–13]. So certain metabolic lipid disorders react to fish oil with the fall in the triglyceride-but not in the apo B level.

The underlying mechanism by which LDL is increased is not clear yet. Relatively dense VLDL poor in triglycerides are metabolized faster to LDL [7]. Fish oil diet leads to a reduction of the VLDL-particle size [12]. Therefore, the VLDL-catabolism is possibly forced and the slight increase of LDL-apo B is explicable. The LDL-clearence or the VLDL-independent LDL synthesis can also be enhanced [11].

The elevation of HDL2-cholesterol can be caused by an increased lipolytic activity.

The glycemic control did not worsen in our study. Other studies with a similar fish oil dosage confirm our finding [11]. In case of a dayly dosage above 4 g n-3 PUFA (eicosapentenoic and docosahexenoic acid) the glycemic control worsens [2, 11].

Summary

In comparison to the rape seed oil supplementation, the fish oil diet effected a decrease in the concentration of serum triglycerides by 29%. LDL-cholesterol increased by 10%, HDL cholesterol by 9% (especially HDL_2-cholesterol), and apo B by 5%. Apo A-I was reduced by 9%. The fasting blood glucose and the glucose tolerance as well as the insulin levels (fasting and after load test) were unchanged at the end of the verum period in comparison to the run-in-phase.

How far the positive changes of the lipoprotein profile (a marked decrease in triglycerides and an increase of HDL-cholesterol) can be neutralized by a mild enhancement of LDL-cholesterol and apo B has to be investigated by further studies.

References

1. Betteridge DJ (1989) Diabetes, lipoprotein metabolism, and atherosclerosis. Brit. Med. Bull. 45:285–311.
2. Friday KE, Childs M, Tsunehora CH, Fujimoto WY, Bierman EL, Ensinck JW (1989) Elevated plasma glucose and lowered triglyceride levels from omega-3 fatty acid supplementation in type II diabetics. Diabet. Care 12:276–283.
3. Herrmann W, Beitz J (1987) Zur Verminderung des atherogenen Risikos durch eikosapentaensäurereiche Diät. Z. ges. inn. Med. 42:117–112.
4. AH Kissebah AH, Alfarsi S, Evans DJ, Adams PW (1982) Integrated regulation of very low density lipoprotein triglyceride and apolipoprotein B kinetics in non-insulin-dependent diabetes mellitus. Diabetes 31:217–225.
5. Kissebah AH (1987) Low density lipoprotein metabolism in noninsulin dependent diabetes mellitus. Diabetes Metab. Rev. 3:619–651.
6. Kostner GM, Karadi I (1988) Lipoprotein alterations in diabetes mellitus. Diabetologica 31:717–722.
7. Packard CJ, Munro A, Lorimer AR, Gotto AM, Shephard J (1984) Metabolism of apolipoprotein B in large triglyceride-rich very low density lipoproteins of normal and hypertriglyceridemic subjects. J. Clin. Invest. 74:2178–2192.
8. Phillipson BE, Rothrock DW, Connor WE, Harris WS, Illingworth DR (1985) Reduction of plasma lipids, lipoproteins and apoproteins by dietary fish oils in patients with hypertriglyceridemia. N. Engl. J. Med. 312:1210–1216.
9. Radack KL, Deck CC, Huster GA (1990) n-3 Fatty acid effects on lipids lipoproteins, and apolipoproteins at very low doses : results of a randomized controlled trial in hypertriglyceridemic subjects. Am. J. Clin. Nutr. 51:599–605.

10. Sanders TA, Sullivan DR, Reeve J, Thompson GR (1985) Triglycerides-lowering effect of marine polyunsaturates in patients with hypertriglyceridemia. Arteriosclerosis 5:459–465.
11. Schectman G, Kaul S, Kissebah AH (1988) Effect of fish oil concentrate on lipoprotein composition in NIDDM. Diabetes 37:1567–1573.
12. Sullivan DR, Sanders TA, Trayner IM, Thompson GR (1986) Paradoxical elevation of LDL apoprotein B level in hyper-triglyceridaemic patients and normal-subjects ingesting fish oil. Atherosclerosis 61:129–134.
13. Taskinen MR, Beltz WF, Harper I, Fields RM, Schonfeld G, Grundy SM, Howard BV (1986) Effects of NIDDM on very-low-density lipoprotein triglyceride and apolipoprotein B metabolism: studies before and after sulfonylurea therapy. Diabetes 35:1268–1277.

Pharmacology of Female Sex Hormones with Respect to Lipid and Lipoprotein Metabolism

A. T. TEICHMANN

As early as the fifties it was apparent that human lipid and lipoprotein patterns showed a marked variation with gender and age (Gofman et al. 1950, Nikkilä 1953, Russ et al. 1951). Attention was drawn to the role of sex steroids in the physiology of lipid and lipoprotein metabolism. Thereafter, it was hypothesized that female sex hormones exert a protective influence on the development of coronary heart disease. Epidemiological surveys such as the Framingham study clearly demonstrated that women of reproductive age experience significantly fewer clinical events due to atherosclerosis as compared to male individuals of the same age. The advantage in coronary morbidity and mortality completely disappears in postmenopausal patients.

Whereas the average differences in lipid and lipoprotein levels between female and male individuals are not significant, the high levels of estrogens and progesterone found during pregnancy have a profound influence (Teichmann et al. 1988). With the onset of the second trimester cholesterol and triglyceride levels rise considerably. Accordingly, the alpha, pre-beta- and betalipoprotein cholesterol levels increase threefold during the second and third trimester as compared to the first.

Pharmacology of Estrogens and Gestagens

The role of exogenous female sex steroids in preventing or promoting atheroma formation by influencing lipid and lipoprotein levels has been widely discussed. The most widespread clinical applications of estrogens and gestagens in primarialy healthy individuals are found in hormonal contraception and postmenopausal replacement. As most women undergo the treatment for several years it seems worthwhile to investigate the pharmacology of hormones used for these purposes.

The estrogens (Table 1) of all modern oral contraceptives contain ethinylestradiol, a derivative of the natural estradiol, that is chemically modified by introducing an ethyl group on carbon 17. This modification leads to a far greater biological activity due to the decreased half-life by inhibition of oxydative metabolism. Estradiol itself is used predominantly in replacement therapy in its mirconized form or conjugated with valerate, succinate and benzoate. Conjugated equine estrogens consist of numerous non-human estrogens extracted from the urine of pregnant mares. Despite the route of ad-

Table 1. Estrogens

Natural		Synthetic
Estradiol	-valerate -succinate -benzoate	Ethinyl-estradiol Mestranol
Estrone Estriol conj. equine Estrogens		

ministration, ethinylestradiol and conjugated estrogens induce the levels of hepatic proteins such as angiotensinogen, sex hormone binding globulin, and corticoid binding globulin up to 5-fold. It ist not clear whether this difference is related to an enhanced biosynthesis or a reduced degradation of these proteins.

As natural progesterone is not suitable for oral medication, due to its very short half-life, a vast number of steoroids have been synthesized with progestogenic activity. According to their chemical structure, these compounds fall into two groups, progesterone derivatives and those of 19-nortestosterone (Table 2). In terms of their biological effects, the secondary hormonal properties of these two groups of progestins are quite different. Whereas substances derived from progesterone exert very little androgenic influences, most of them have moderate or marked anti-androgenic effects resulting from non-functional binding to the androgen receptor. In contrast, the derivatives of 19-nortestosterone have androgenic properties, namely those belonging to the subgroup of 13-ethyl-gonanes like norgestrel, desogestrel, gestoden and norgestimate. In comparison to the progesterone-like progestins they do not show any glucocorticoid activities (Table 3).

When comparing the effects of different estrogens and gestagens, it is important to note that in addition to the chemical structure of a steroid, a variety of pharmacological as well as pharmacokinetic parameters can substantially modify the metabolic result of the steroid's action in vivo. Table 4 summarizes the factors that modify the biological effects of sex steroids. Thus not only chemical structure and dosage of a steroid has to be considered but also the route of administration leading to the first passage in the liver in the case of oral intake, the induction or inhibition of receptor formation as a result of duration of use or interference with other hormones, the pretherapeutic hormonal status of the patient and her metabolic conditions. When considering all of these factors, the effects of hormones of the same pharmacological group can vary widely and sometimes appear paradoxical. For example the oral intake of estrogens can lead to a rise in blood pressure as observed with ethinylestradiol in combination with a gestagen for contraceptive purposes; in contrast, estradiol taken orally by postmenopausal subjects is able to exert antihypertensive effects. Moreover, the formation of gestagen receptors depends on the presence of an estrogenic stimulus, whereas gestagens downregulate their own receptors by entire antiestrogenic properties and thus lead

Table 2. Gestagens

Progesterone derviates	*19-Nortestosterone-derivatives*
Medroxyprogesteroneacetate	*Estranes (β-methyl-gonanes)*
Mergestrolacetate	Norethisterone
Medrogestone	Norethisteroneacetate
Chlormadinoneacetate	Ethynodioldiacetate
Cyproteroneacetate	Norethynodrel
	13-Ethyl-gonanes
	D-Norgestrel
	Desogestrel
	Gestoden
	Norgestimate

Table 3. Secondary hormonal properties in the rat

Progestoges	Antigonadotrophic activity[a]	Estrogenicity[b]	Antiestrogenicity[c]	Androgenicity[d]	Antiandrogenicity[e]	Glucocorticoid[f]
Pro	+	0	+	0	+	+
LNG	+++	0	+++	+++	0	0
MPA	+++	0	+	+	+	++
NET	++	+	++	++	0	0
NEA	++	+	++	++	0	0
NED	++	++	0	+	0	0
EDA	++	+	+	+	0	0
LYN	++	+	++	++	0	0
CMA	++	0	++	0	++	+
CPA	+++	0	++	0	+++	+
MGA	++	0	+	0	+	+
HPC	+	0	+	0	+	+

Table 4. Biological effects of sex hormones – pharmacological determinants

- Chemical structure
  receptor affinity
  intrinsic activity
  half life
  metabolites
  protein binding
- Dosage
- Route of administration
- Duration of use
- Interference with other hormones and drugs
- Patient's endocrinological condition
- Patient's metabolic conditions
  (enzyme activities, diseases as hypertension, diabetes, dyslipoproteinaemia ...)

to limitation of their own biological action. Furthermore, serum concentrations of ethinylestradiol and all synthetic gestagens show an up to 20-fold inter- and intraindividual variation when administered in a standardized manner. In summary, metabolic effects of female sex hormones should be regarded as the results of a highly complicated system. Therefore, analogies between endogenous and exogenous hormones or between the actions of different hormones in different groups of women should be drawn with caution.

Exogenous Sex Hormones and Lipoprotein Metabolism

Changes in lipid and lipoprotein concentrations induced by estrogens are well documented in the literature.

In cases of oral administration, the effects are much more pronounced due to the first passage effect in the liver. There are three main metabolic pathways leading to the characteristic modifications of lipid and lipoprotein concentrations. First, the ability of estrogens to stimulate protein synthesis leads to apoprotein synthesis (I. E. Archer et al. 1986, Tam et al. 1985, Farish et al. 1984, 1986). Moreover, kinetic studies with radiolabelled HDL particles (Hazzard et al. 1984) clearly demonstrated that estrogens slow down the HDL_2 turnover. This latter finding corresponds very well to the studies of Tikkanen et al. 1980, who describe the hormone dependency of hepatic lipase, the key enzyme in HDL-metabolism. (Appelbaum et al. 1977, Tikkanen et al. 1982). The well-known fact that the LDL cholesterol levels fall in postmenopausal women undergoing estrogen replacement may be explained by the following animal experiments. Kovanen et al. (1979) and Windler et al. (1980) demonstrated that apo B, E receptors are inducible by estrogens in rats. Triglyceride and VLDL levels are positively influenced by estrogens. This effect may be due to an elevated triglyceride turnover rate (Glueck and Fallart 1974) and a stimulated secretion rate of VLDL and triglycerides in the liver (Glueck et al. 1975, Kekki und Nikkila 1971). It is not clear whether estrogens interfere with the VLDL catabolism. The main and well documented effects of estrogens are summarized in Table 5 (Figs. 1–5).

The role of progestins in lipo- and lipoprotein metabolism is more complicated than the role of estrogens. The heterogeneity of synthetic hormones exerting progestogenic activities is responsible for confounding results in the literature. As shown in Figs. 6–12, the derivatives of 19-nortestosterone induce far more pronounced changes in lipo- and lipoprotein levels than the typi-

Table 5. Effects of estrogens on LP-metabolism

Metabolic pathway		Serum concentration	
VLDL-Synthesis	↑	VLDL	↑
Hepatic LDL receptors	↑	LDL	↓
Hepatic lipase	↓	HDL_2	↑
Hepatic HDL-synthesis	↑	HDL_3	↑

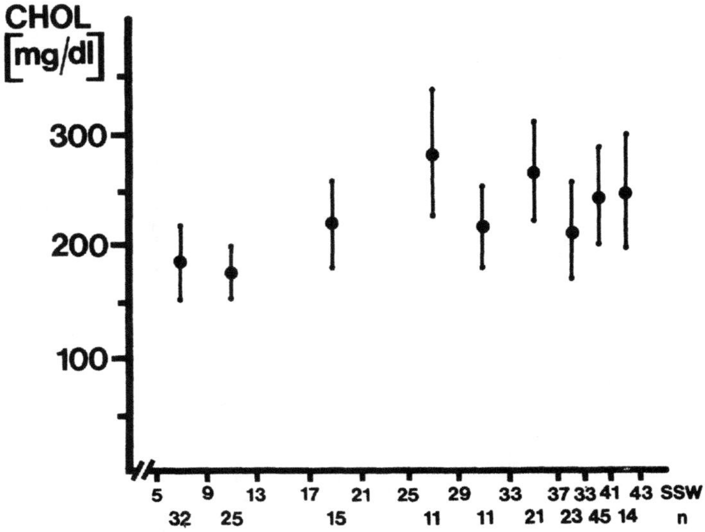

Fig. 1. Cholesterol levels in pregnant patients (medians, upper and lower limits) (Teichmann et al. 1988b.) *SSW* = weeks of gestation; *n* = number of subject

cal progesterone derivative, medroxyprogesterone acetate. However, in a dose dependent manner high amounts of medroxyprogesterone acetate given orally do lead to the typical gestagenic pattern also observed with androgens. (Teichmann et al. 1987, 1988). The mechanism of action of gestagens is related to their anti-estrogenic properties. Moreover, gestagens seem to have a

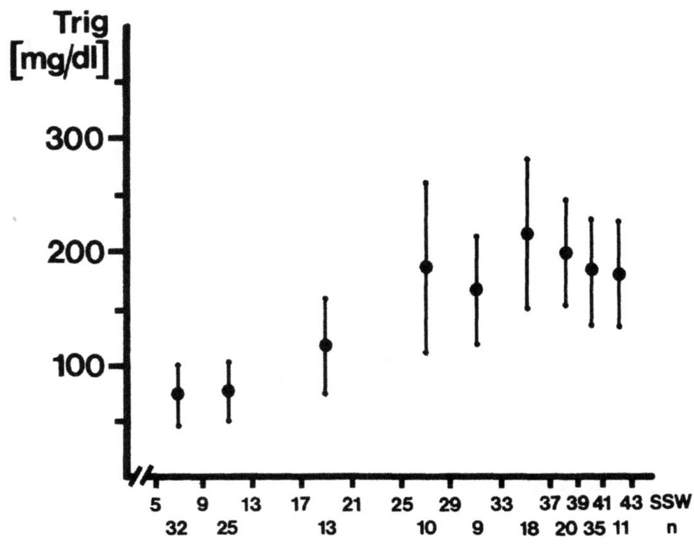

Fig. 2. Triglycerone levels in pregnant patients (medians, upper and lower limits) (Teichmann et al. 1988b.) *SSW* = weeks of gestation; *n* = number of subject

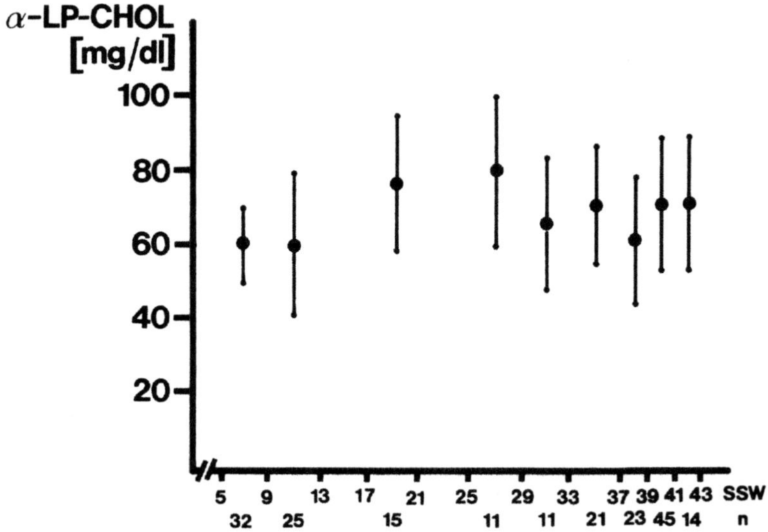

Fig. 3. α-Lipoprotein cholesterol levels in pregnant patients (medians, upper and lower limits) (Teichmann et al. 1988b.) *SSW* = weeks of gestation; *n* = number of subject

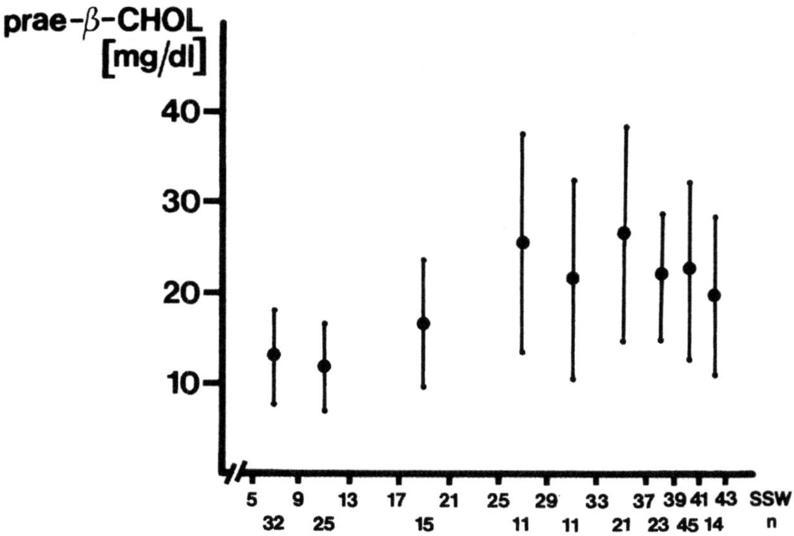

Fig. 4. pre-β-lipoprotein cholesterol levels in pregnant patients (medians, upper and lower limits) (Teichmann et al. 1988b.) *SSW* = weeks of gestation; *n* = number of subject

Table 6. Effects of gastagens on LP-metabolism

Metabolic pathway		Serum concentration	
VLDL-Secretion	↓	VLDL	↓
LDL	?	LDL	(↑)
Hepatic lipase	↑	HDL_2	↓

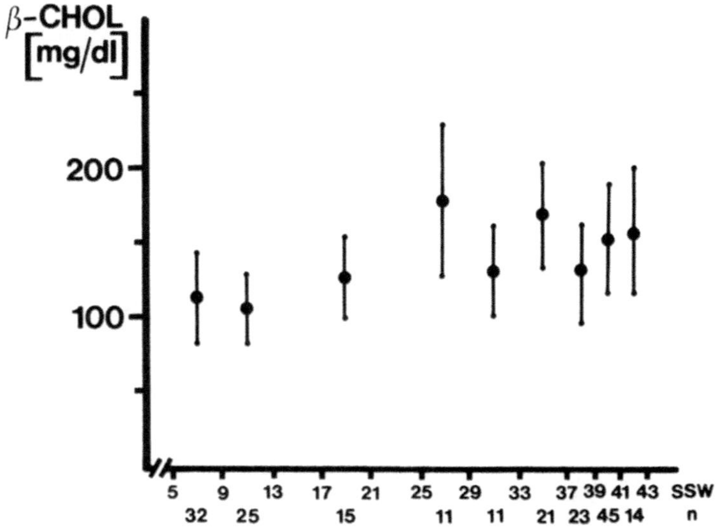

Fig. 5. β-Lipoprotein cholesterol levels in pregnant patients (medians, upper and lower limits) (Teichmann et al. 1988b.) *SSW* = weeks of gestation; *n* = number of subject

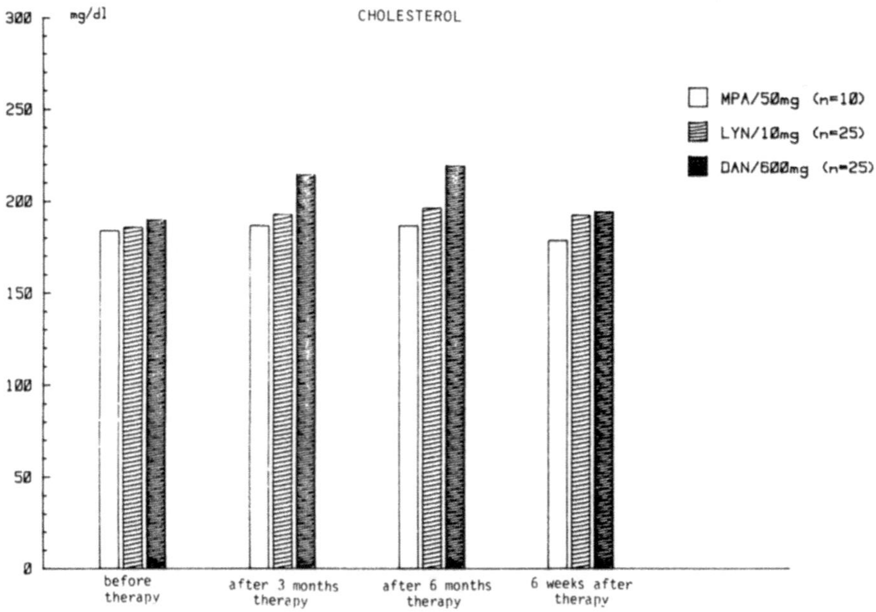

Fig. 6. Lipid and lipoprotein levels (medians) in patients with endometriosis before during and after treatment with different steroids (medroxyprogesteroneacetate, lynestrenol, danzol) (Teichmann et al. 1988b)

direct influence on the activity of the hepatic lipase (Nikkila et al. 1980 and Tikkanen et al. 1980). Table 6 summarizes the net effects and metabolic pathways of the influence of progestins on lipid- and lipoprotein levels.

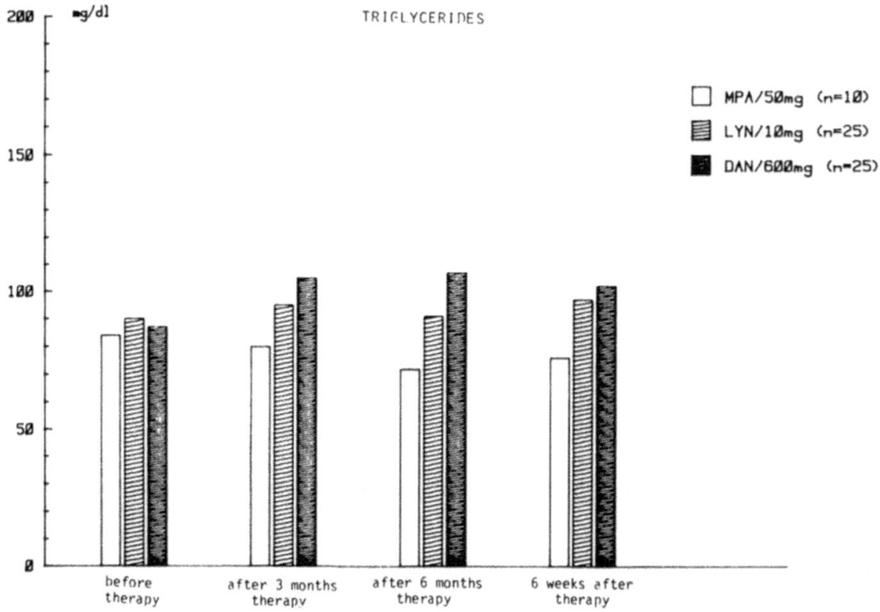

Fig. 7. Lipid and lipoprotein levels (medians) in patients with endometriosis before during and after treatment with different steroids (medroxyprogesteroneacetate, lynestrenol, danzol) (Teichmann et al. 1988b)

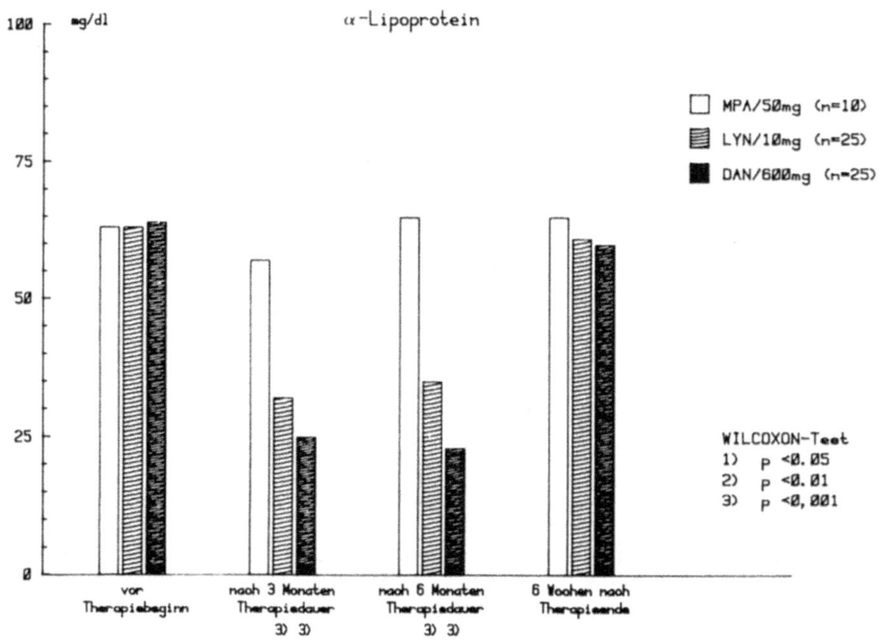

Fig. 8. Lipid and lipoprotein levels (medians) in patients with endometriosis before during and after treatment with different steroids (medroxyprogesteroneacetate, lynestrenol, danzol) (Teichmann et al. 1988b)

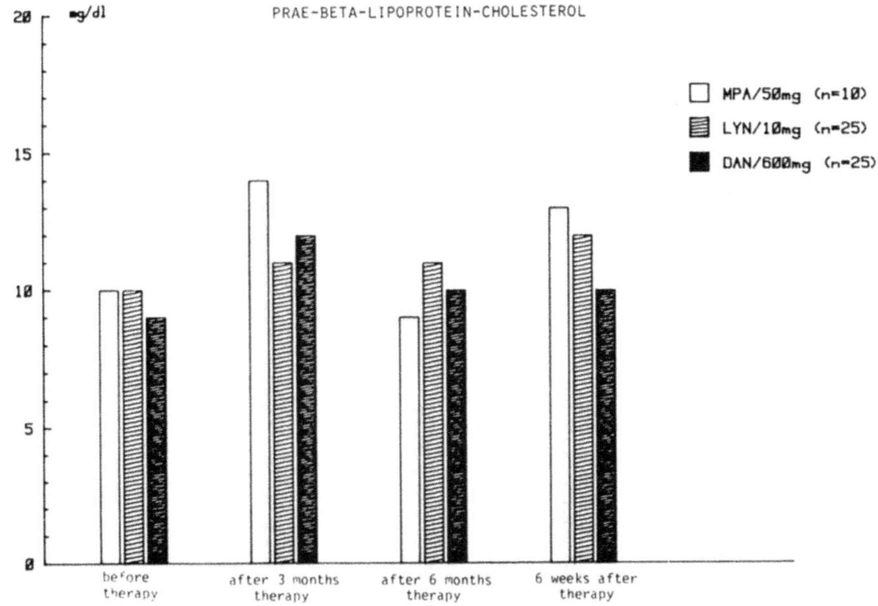

Fig. 9. Lipid and lipoprotein levels (medians) in patients with endometriosis before during and after treatment with different steroids (medroxyprogesteroneacetate, lynestrenol, danzol) (Teichmann et al. 1988b)

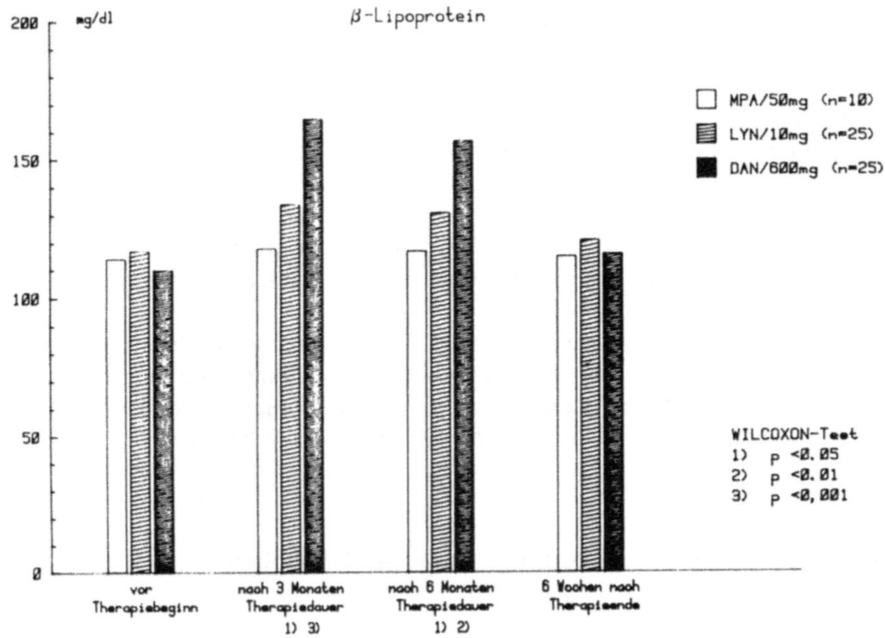

Fig. 10. Lipid and lipoprotein levels (medians) in patients with endometriosis before during and after treatment with different steroids (medroxyprogesteroneacetate, lynestrenol, danzol) (Teichmann et al. 1988b)

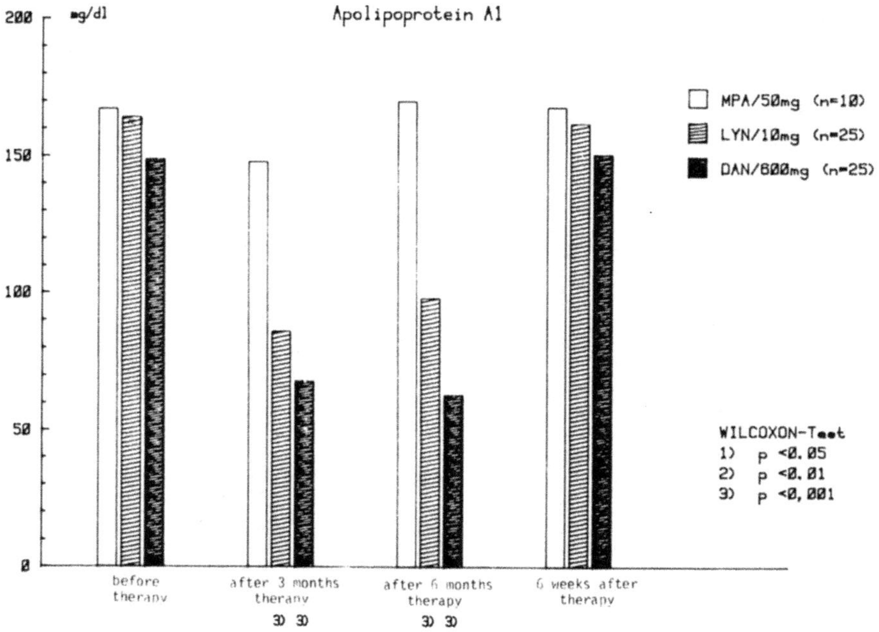

Fig. 11. Lipid and lipoprotein levels (medians) in patients with endometriosis before during and after treatment with different steroids (medroxyprogesteroneacetate, lynestrenol, danzol) (Teichmann et al. 1988b)

In practice, most patients are treated with a combination of female sex hormones whether in contraceptive preparations or in combinations for replacement therapy. The resulting lipid- and lipoprotein profiles depend on the dosages and on the chemical characteristics of the components used. Table 7 gives a broad overview of studies with oral contraceptives. In summary, low dose preparations do not show any major net effect on the cholesterol, HDL or LDL concentrations, whereas the levels of triglycerides and VLDL rise slightly upon treatment.

In contrast to hormonal contraceptives, preparations for replacement therapy which do not contain a predominant gestagenic component, generally exert the beneficial effect of estrogens on LDL cholesterol levels (Table 8). HDL levels are found to be elevated with these combinations.

Exogenous Sex Hormones, Lipoprotein Metabolism and Atherosclerosis

Numerous epidemiological studies have demonstrated that death due to circulatory diseases far exceeds all other mortality factors, in particular malignant diseases. Female sex hormones and vascular risk are determined mainly

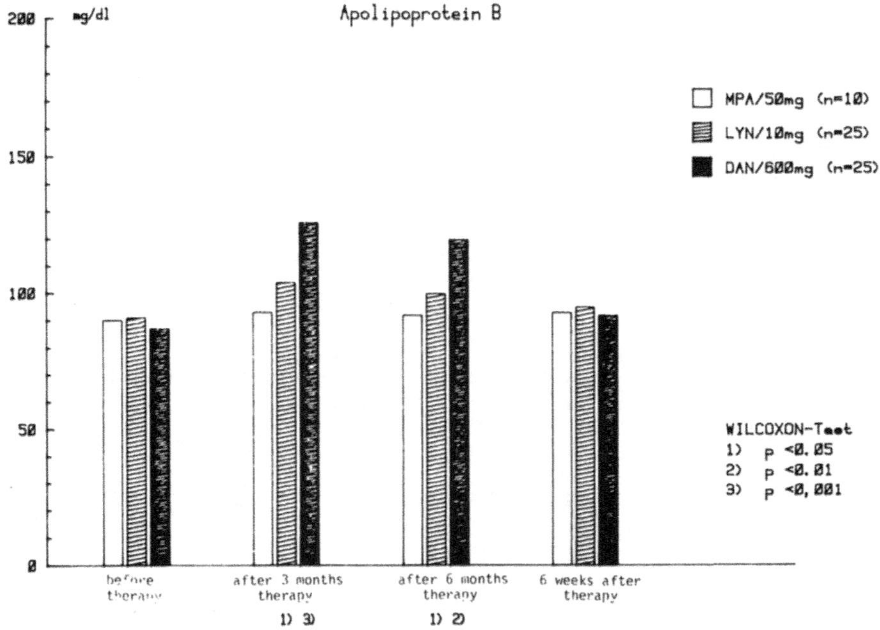

Fig. 12. Lipid and lipoprotein levels (medians) in patients with endometriosis before during and after treatment with different steroids (medroxyprogesteroneacetate, lynestrenol, danzol) (Teichmann et al. 1988b)

by influences on the hemostatic system: blood pressure, carbohydrate metabolism and finally influences on lipid and lipoprotein metabolism. Direct actions of hormones on the venous and arterial vessel wall have yet to be proven. Data concerning lipid and lipoproteins with respect to cardiovascular morbidity and mortality were generated in the Framingham study (Castelli et al. 1986). This epidemiological survey conducted on postmenopausal women irrespective of replacement therapy clearly showed that HDL cholesterol concentrations are the most powerful predictors for death by cardiovascular diseases. When, in 1982, results of the oral contraceptive study conducted by the Royal College of General Practitioners (Kay 1982) were published indicating that death rates by arterial diseases rose with dose of the gestagenic component, these findings were correlated with the HDL cholesterol levels reduced by the according preparations (Fig. 13). It was concluded that higher death rates were due to a decrease in HDL cholesterol concentrations. Thereafter it became accepted that female sex hormones, and gestagens in particular, may lead to atherosclerosis by influencing the HDL metabolism. It was widely neglected that no persisting mortality by cardiovascular diseases existed in ex-users of the birth control pill, which one should have expected, if atherosclerosis was induced by intake of oral contraceptives. Moreover it was widely ignored that autopsies on young women who had been using oral contraceptives and who had died as a result of myocardial infarction, very ra-

Table 7. Effects of combined oral contraceptives on lipid and lipoprotein levels

Autors	Year	Gestagen	Oestrogen	Chol	VLDL-C	LDL-C	HDL-C	Trig.	n
Krauss et al.	1983	Norethisteronc 400 µg	35 µg	+13%	0	0	+10%	0	10
Rössner et al.	1982	Norethisterone 500 µg	35 µg	+ 7%	0	0	0	0	10
Rössner et al.	1982	Norethisterone 1 mg	35 µg	0	0	+12%	– 6,2%	0	11
Rössner et al.	1982	Norethisterone 1 mg	50 µg	+ 8%	0	+15%	– 6%	0	11
Wahl et al.	1983	Norethisterone 1/2 mg	50/80/100 µg	+10%	+33%	0	+ 7%	+57%	146
Wahl et al.	1983	Norethisterone-ac 1/2,5 mg	50 µg	+10%	0	+14%		+56%	41
Wahl et al.	1983	Ethynodiol-diac. 1 mg	50 µg	+19%	+82%	0	0	+78%	17
Wahl et al.	1983	Ethynodiol-diac. 1 mg	100 µg	+ 9,2%	+64%	+14%	0	+96%	64
Wahl et al.	1983	Norethynodrel 2,5/5 mg	75–100 µg	0	0	0	+35%	+83,3%	12
Crona et al.	1982	Lynestrenol 0,2/5 mg	50 µg	0	0	0	0	+35%	12
Crona et al.	1982	D-Nurgestrel 50/125 µg	50 µg	0	0	0	0	+35%	12
Rössner et al.	1980	D-Norgestrel 150 µg	30 µg	0	0	0	–14%	0	11
Ahren el al.	1981	D-Norgestrel 150 µg	30 µg	0	0	0	0	0	18
Mendoza et al.	1981	D-Norgestrel 150 µg	30 µg	+12%		+12%	0	+34%	30
Taggart et al.	1981	D-Norgestrel 150 µg	30 µg	+ 6%	0	0	0	0	15
Samsioc	1982	D-Norgestrel 150 µg	30 µg	0	+36%	+11%	– 9%	+41%	7
van der Sleeg el al.	1977	D-Norgestrel 250 µg	50 µg	0	0	+20%	0		26
Rössner et al .	1980	D-Norgestrel 250 µg	50 µg	0	0	0	–11%	0	11
Mendoza et al.	1981	D-Norgestrel 250 µg	50 µg	+13%		+12%	0	+50%	24
Mendoza et al.	198I	D-Norgestrel 500 µg	50 µg	0		0	0	0	17
Wahl et al.	1983	D-Norgestrel 300 µg	50 µg	+11%	+42%	+18%	+18%	+40%	76
Krauss et al.	1983	D/L-Norgestrel 300 µg	30 µg	0		0	0	+18%	11
Samsioe	1982	Desogestrel 150 µg	30 µg	0	+23%	0	+32%	+21%	8
März et al.	1985	Desogestrel 150 µg	30 µg	0	0	0	+10%	+21%	11
Klosterboer et al.	1986	Desogestrel 150 µg	30 µg	0	0	0	0	+35%	16
Jung-Hoffman et al.	1987	Desogestrel150 µg	30 µg	0	+30%	0	+10%	+20%	10
Jung-Hoffmann et al.	1987	Gestoden 75 µg	30 µg	0	+30%	0	+10%	+20%	10
Larsson-Cohn et al.	1986	Cyproteronaz. 2 mg	30 µg	0	+82%	0	+14%	+74%	22
Cirkel et al.	1986	Chlormadinonaz. 1/2 mg	50 µg	+13%		0	+27%	+65%	22

Table 8. Lipid and lipoprotein levels in patients on combined hormonal replacement therapy

Authors	Year	Dosage	Oestrogen	Period	Dosage	Gesta-gens	Period	Chol.	Trig.	VLDL	LDL	HDL	LDL/HDL	n
Ottoson et al.	1985	2 mg	Estradiol-val.	21 d	100 mg	prog +	10 d					0		18
Mattson et al.	1982	1,25 mg	Konj. oestrogens	21 d,	5 mg	MPA	5 d	0	0	0	0	+		8
Nachtigall et al .	1979	2,5 mg	Konj. oestrogens	21 d,	10 mg	MPA	7 d						+ [a]	84
Hirvonen et al.	1981	2 mg	Estradiol-val.	24 d,	10 mg	MPA	10 d	–			0	0		6
Ottoson et al.	1985	2 mg	Estradiol-val.	21 d,	10 mg	MPA	10 d					-HDL_2		20
Notelovitz et al.	1983	0,625[c] 125 mg	Konj. oestrogens	21 d,	10 mg	MPA	7 d				–	+	–	20
Volpe et al.	1983	0,625 mg	Konj. oestrogens	24 d,	10 mg	MPA	10 d	0	0	0	0	0		10
Tikkanen et al.	1982a	2 mg	Estradiol-val.	28 d,	150 μg	NG	15 d	0	–	0	0	-HDL_2		11
Ottoson et al.	1985	2 mg	Estradiol-val.	21 d,	250 μg	NG	10 d					-HDL_2		20
Paterson et al.	1980	2 mg	Estradiol-val.	21 d,	500 μg	NG	10 d	–	+					11
Hirvonen et al.	1981	2 mg	Estradiol-val.	24 d,	500 μg	NG	10 d	–[b]			0			6
Hirvonen et al.	1981	2 mg	Estradiol-val.	24 d,	10 mg	NET	10 d	–[b]			0	–		6
Volpe et al.	1983	0,625 μg	Konj. oestrogens	24 d,	10 mg	NET	10 d	0	0	0	+	–		10
Mattsonetal.	1984	2/1 mg	1 E2/E3cont.		1 mg	NEA	cont.	–	0	+	~	–	0	26

[a] only if LDL/HDL ratio was elerated before treatment
[b] most pronounced cholesterol decrease in hypercholesterolemic pretreatment values
[c] micronized form

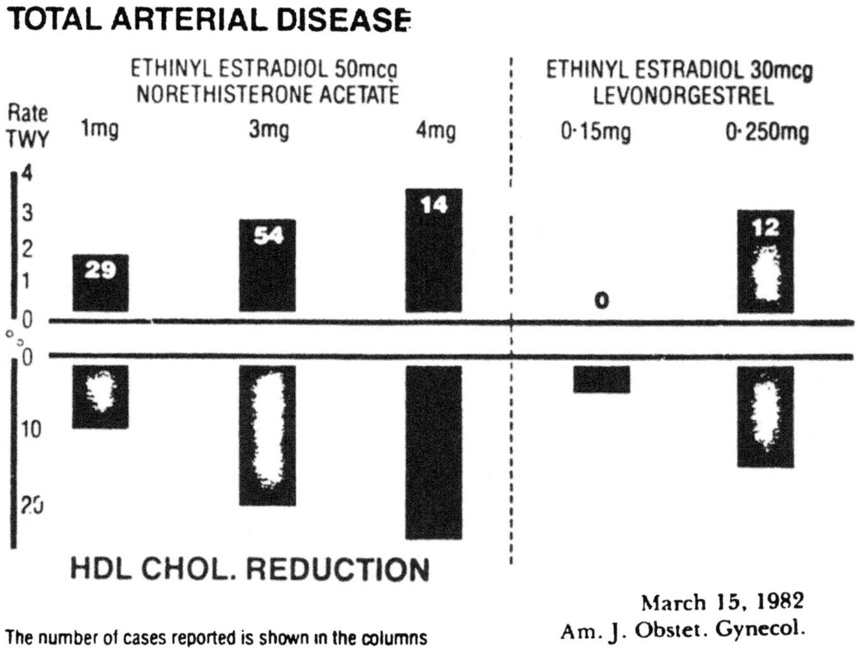

Fig. 13. Dose of gestagen, HDL-cholesterol levels and rate of arterial diseases in patients taking combined oral contraceptives (Kay 1982)

rely showed atherosclerotic lesions, but in the vast majority the infarction was caused by thrombotic events.

An analysis of the pathophysiological model of HDL turnover (Fig. 14) suggests that the "HDL hypothesis" with respect to all contraceptives must be revised. As mentioned above, gestagens stimulate one of the key enzymes of the HDL metabolism, the hepatic lipase. Stimulating hepatic lipase enhances HDL metabolism by accelerating the degradation of cholesterol rich HDL2 particles. This may represent a more effective reverse cholesterol transport which is believed to be one of the most important metabolic steps in removing cholesterol from the circulatory system. Even if other factors related to HDL may contribute to its anti-atherogenic effects, it is clear that simple conclusions based on levels of peripheral parameters do not reflect the dynamic metabolic situation, which seems to be far more important and also far more complicated.

Clarkson and coworkers conducted a most interesting experiment with Cynomolgus monkeys (1990). These monkeys display very similar lipoprotein profiles and changes as do women taking hormonal contraceptives. The animals were fed an atherogenic diet and then treated with an oral contraceptive (Ovral) as well as with a vaginally applied estradiol levonorgestrel preparation. As would be expected, the resulting lipid and lipoprotein profiles showed a marked fall in HDL cholesterol levels and raised LDL cholesterol concentrations in women taking the oral contraceptive. Surprisingly, the

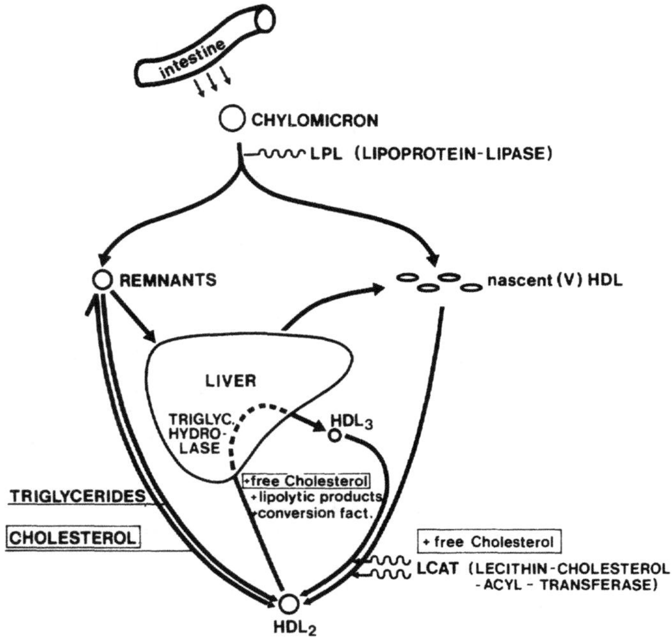

Fig. 14. HDL-mediated reverse cholesterol transport (mod. acc. Teichmann et al. 1989)

autopsic findings clearly showed that the animals with the "worst" lipoprotein profile that had been administered the oral contraceptive presented only minimal atherosclerotic plaques whereas the most extensive atherosclerotic lesions were found in the animals treated by the vaginal route.

In summary, it can be concluded that in oral contraceptive users lipid- and lipoprotein changes do not lead to atherosclerosis and thus should neither be regarded as atherogenic nor anti-atherogenic. An analogy between different lipoprotein profiles induced by sex hormones in women of reproductive age with those women not treated with hormones and of postmenopausal age should not be drawn! In contrast, the epidemiological studies conducted thus far, do not leave any doubt that hormone replacement therapy reduces the risk of cardiovascular mortality in postmenopausal women. The property of estrogens to lower LDL levels in estrogen-deprived individuals by inducing formation of apo-B-receptors may be the main metabolic effect by which at least atherosclerotic complications may be prevented to a certain extent.

References

Ahren T, Lithell H, Victor A, Vessby B, Johannson EDB (1981) Comparison of the metabolic effects ot two hormonal contraceptive methods: an oral formulation and a vaginal ring. II. Serum lipoproteins and apolipoproteins. Contraception 24: 451–468.

Archer TK, Tam SP, Deeley RG (1986) Kinetics of estrogen-dependent modulation of apolipoprotein A-I synthesis in human hepatoma cells. J. Biol. Chem. 261: 5067–5074.
Castelli WP, Garrison RJ, Wilson P, Assert RD, Kaldusian S, Kamal WB (1986) Incidence of coronary heart disease and lipoprotein cholesterol levels. The Framingham study. J. Am. Med. Ass. 256:2835–2838.
Clarkson TB, Shively CA, Morgan TM, Koritnik DR, Adams MR, Kaplan JR (1990) Oral contraceptives and coronary artery atherosclerosis of cynomolgus monkeys. Obstet. Gynecol. 75:217–222.
Farish E, Fletcher CD, Hart DM, Teo HT, Allazawi F, Howie C (1986) The effect of conjugated equine oestrogens with and without a cyclial progestogen on lipoproteins, and HDL subfractions in postmenopausal women. Acta Endocrinol. 113: 123–127.
Farish E, Fletcher CD, Hart DM, Allazawi F, Abdalla HI Gray CE (1984) The effects of hormone implants on serum lipoproteins and steroid hormones in bilaterally oophorectomised women. Acta Endocrinol. 106:116–120.
Glueck CJ, Fallat R (1974) Gonadal hormones and triglyzerides. Proc. R. Soc. Med. 67:667–669.
Glueck CJ, Fallat R, Scheel D (1975) Effects of estrogenic components on triglyzeride kinetics. Metabolism 24:537–545.
Gofman JW, Jones HB, Lindgren FT, Lion TP, Elliot HA, Strisower B (1950) Blood lipids and human atherosclerosis. Circulation 2:161–178.
Hazzard WR, Haffner SM, Kushawaha RS, Appelbaum-Bowden D, Foster DM (1984) Preliminary report: kinetic studies on the modulation of high density lipoprotein, apolipoprotein, and subfraction metabolism by sex steroids in a postmenopausal woman. Metabolism 33:779–784.
Kay CR (1982) Progestogens and arterial disease. J. Am. Obstet. Gynec. 142:762.
Kekki M, Nikkila EA (1971) Plasma triglyceride turnover during use of oral contraceptives. Metabolism 20:878–889.
Kovanen PT, Brown MS, Goldstein JL (1979) Increased binding of low density lipoprotein with 17 A-ethinyl estradiol. J. bio. Chem. 254:11367–11373.
Nikkila EA (1953) Studies on the lipid-protein relationship in normal and pathological sera and the effect of heparin on serum lipoproteins. Scand. J. Clin. Lab. Invest. 5:1–101.
Nikkila EA, Tikkanen M, Sipinen S (1980) Decrease of plama high density lipoprotein (HDL) and increase of postheparin plasma hepatic lipase activity during progestin (levonorgestrel) treatment. Circulation 62:228.
Russ EM, Eder HA, Barr DP (1951) Protein-lipid relationship in human plasma. I. In normal individuals. Amer. J. Med. 11:468–479.
Tam SP, Archer TK, Deeley RG (1985) Effects of estrogen on apolipoprotein secretion by the human hepatocarcinoma cell line, HepG2. J. Biol. Chem. 260:1670–1675.
Teichmann AT, Cremer P, Wieland H, Kuhn W, Seidel D (1988a) Lipid metabolic changes in the hormonal treatment of endometriosis. Maturitas 10: 27–33.
Teichmann AT, Wander HE, Cremer P, Wieland H, Kuhn W, Nagel GA, Seidel D (1987a) Medroxyprogestrone acetate and lipid metabolic changes. Z. fur Arzneimittelforschung/Drug Research.
Teichmann AT, Wieland H, Cremer P, Kulow G, Mehle U (1988b) Serumlipid- und Lipoproteinkonzentrationen in der Schwangerschaft und zum Zeitpunkt der Geburt bei normalen sowie durch hypertensive Gestose und kindliche Mangelentwicklung komplizierten Schwangerschaftsverlauf. Geburtsh u Frauenheilk 48:134.
Teichmann AT (1991) Kontrazeption. Wissenschaftliche Verlagsgesellschaft, Stuttgart.
Teichmann AT, Wieland H, Brockerhoff P (1989) Sexualhormone, Fettstoffwechsel und Arteriosklerose der Frau. Wissenschaftliche Verlagsgesellschaft, Stuttgart.
Tikkanen MJ, Kuusi T, Sipinen S, Nikkila EA (1980) Hepatic lipase and plasma HDL concentrations during progestogen treatment in women. Eur. J. Clin. Invest. 10:37.

Tikkanen MJ, Nikkila EA, Kuusi T, Sipinen S (1982b) High density lipoprotem - 2 and hepatic lipase: Reciprocal changes produced by estrogen and norgestrel. J. Clin. Endocr. 54:1113–1117.
Windler EE, Kovanen PT, Chao YS, Brown MS, Havel RJ, Goldstein JL (1980) The estradiol-stimulated lipoprotein receptor of rat liver: A binding site that mediates the uptake of rat lipoprotein containlng apoproteins B and E. J. Biol. Chem. 255: 10464.

Why Women Live Longer than Men (or Why Can't a Man be More Like a Woman?)

W. R. HAZZARD

The changing demography of an aging population common to all Western nations results in a shift of the population distribution by age from the triangular configuration characteristic of developing nations (with high birth rates, high infant mortality, and the highest proportion of the population in the youngest age group) to the rectangular distribution characteristic of developed nations wherein relatively equal proportions are distributed across the age range until beyond age 75 (Fig. 1) [1]. Given that mortality rates in women are lower than men in all developed nations across the entire life span, this "rectangularization" of the population distribution by age results in a progressive predominance of women with increasing age. This produces a gender ratio of survivors which increases exponentially, especially beyond age 60 such that in the ninth and tenth decades there are nearly three times as many women as men. This predominance of women is further exaggerated in the long term care arena, in which residential status is a function of survival, dependence, and lack of a surviving spouse; hence here this Ratio is even higher (e.g., 12 to 1 in the North Carolina Baptist Homes for the Aging).

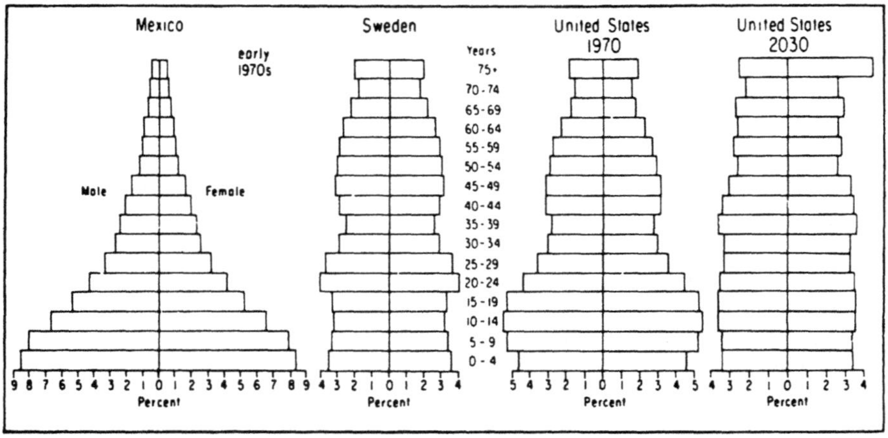

Fig. 1. Comparative population profiles: age and sex distribution. Reprinted with permission from Bureau of the Census, National Center for Health Statistics and United Nations Demographic Yearbook, from Somers AR (1978) The high cost of health care for the elderly: Diagnosis, prognosis, and some suggestions for therapy. Journal of Health Politics, Policy and Law 3:163-180. Copyright, Department of Health Administration, Duke University, Durham, NC. Reprinted with permission

Inspection of the gender ratio and mortality rates across the life span displays a "spike and dome" configuration (Fig. 2) [2], suggesting at least two groups of causes for the greater survival of women over men. Further insight as to the cause of this disparity is provided by reference to a table of gender ratios in the leading causes of death (Table 1) [3]. Here it can be appreciated that there are those causes of male mortality peaking at ca. age 20 (accidents, homicide, suicide, all related to substance abuse and risk-taking behaviour) and those attributed to the chronic diseases to which men are at greater risk, led by heart disease but including also all other major causes of death (several of which are also attributable to substance abuse, notably cigarette smoking and alcohol). The single exception to the substantially greater male mortality rates is diabetes mellitus, in which women are equally predisposed to death as are their male diabetic counterparts. In the aggregate the sex differential in coronary heart disease accounts for the greatest proportion (ca. 40–50%) of the sex differential in longevity [4], though the other major causes (e.g., lung cancer, respiratory diseases, and other malignancies) share in common with coronary heart disease the central role of cigarette abuse in their pathogenesis. And indeed the historical sex differential in cigarette smoking, men starting earlier and smoking heavier than women across the entire adult life span (a differential which is narrowing rapidly in the late 20th century in the United States), is thought to represent a major contributor to the sex differential in longevity, accounting for as many as four of the seven years of greater longevity at birth enjoyed by women compared to men in Western societies [5].

Insights as to the role of sex hormones in the sex differential in CHD is provided by reference to the sex ratio in CHD mortality across the adult life span: whereas overall CHD mortality is twice as great in men as in women, it is as high as 5–6 fold higher in men than women prior to the years of the men-

Table 1. Male/female ratio of age-adjusted death rate for the ten leading causes of death, united states, 1979, all races combined*

Causes of Death by Rank	Male/female ratio (%)
Heart disease	199
Malignant neoplasms	151
Cerebrovascular diseases	119
All accidents	293
Chronic obstructive lung disease	293
Pneumonia and influenza	177
Diabetes mellitus	102
Cheronic liver disease	216
Atherosclerosis	132
Suicide	133
All causes	179

* Rank based on number of deaths from each cause in 1980.
Printed with permission from Wylie CM: Contrasts in the health of elderly men and women: An analysis of recent data for whites in the United States. JAGS 32:670, 1984.

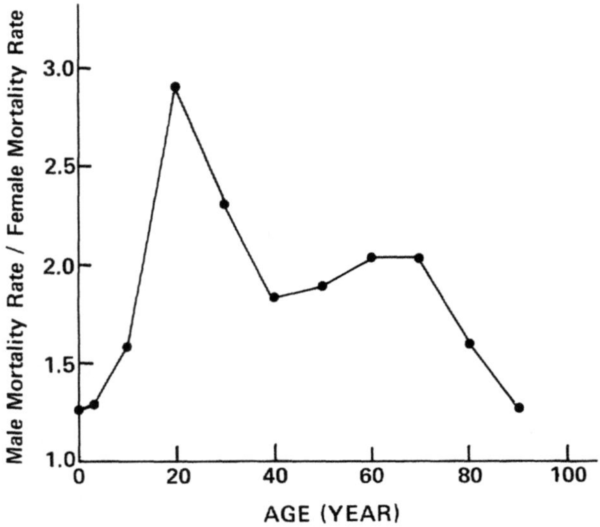

Fig. 2. Sex mortality ratio, US, 1976. Compiled from data presented in Gee EM, Veevers JE: Accelerating sex differentials in mortality: An analysis of contributing factors. Social Biology 30:75-85, 1984. Reprinted with permission from Hazzard WR (1986) Biological basis of the sex differential in longevity. JAGS 34:455-471

opause, the gender ratio declining progressively thereafter to being just over unity in the eighth and ninth decades (CHD risk is never higher at any given age in women than men). Moreover, in the perimenopausal years the CHD rate is substantially higher in postmenopausal than premenopausal women of comparable age [6]. Hence it appears likely that relative immunity to CHD is conferred upon premenopausal women by virtue of their hormonal status. This suggestion is substantiated by reference to distributions of lipoprotein lipid levels beyond puberty in American population studies (Fig. 3) [7]. Studies such as those of the Lipid Research Clinics of North America in the 1970's disclosed that low density lipoprotein (LDL), high density lipoprotein (HDL) and LDL/HDL cholesterol levels are equivalent in males and females until puberty. Between puberty and menopause, however, LDL levels rise faster and remain higher in men than in women, this pattern being reversed across the menopause such that postmenopausal women have higher average LDL cholesterol levels than their male counterparts of comparable age. HDL levels decline dramatically in boys across puberty (and in parallel with adolescent development on the Tanner scale) and remain lower in men than in women throughout the remainder of the life span. This lower average HDL level in males than females exceeds the differential in LDL; hence, LDL/HDL ratios are higher in men than in women at all points in the adult life span beyond puberty, but the rise in LDL across the menopause in women (presumably attributable to estrogen withdrawal) results in a declining LDL/HDL gender ratio in the postmenopausal era. Thus it is tempting to hypothesize that it is the sex differential in sex hormones that confers the sex differential in lipoprotein metabolism which, given a common Occidental lifestyle, produces the sex differential in atherosclerosis and this accounts for the sex differential in longevity. Overall this would appear a plausible but in-

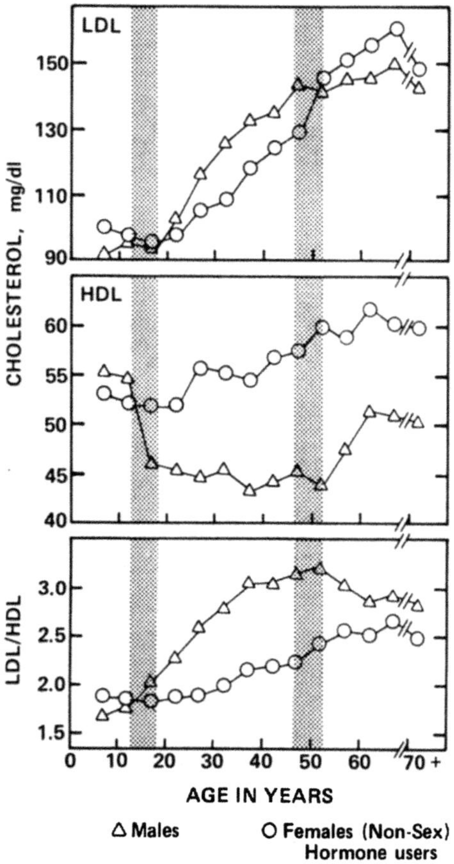

Fig. 3. Gender differences in serum cholesterol levels. The figure includes the levels of low density lipoprotein (LDL) cholesterol (top) in men (triangles) and in women (circles); the levels of high density lipoprotein (HDL) cholesterol (middle); and the LDL/HDL ratio (bottom). Plotted are the average values found in the North American population by the Lipid Research Clinic Prevalence Survey. Reprinted with permission from Hazzard WR. A central role of sex hormones in the sex differential in lipoprotein metabolism, atherosclerosis, and longevity. In: Ory MG, Warner HR, ed. (1990) Gender, health, and longevity: multidisciplinary perspectives. New York: Springer Publishing Company, 87–108

complete explanation, the sex differential in lipoproteins accounting for 25–50% of the sex differential in CHD [4]) (and hence ca. 10–25% of the total sex differential in longevity).

Epidemiological support for an important role for sex hormones in determining lipoprotein lipid levels (and atherogenic risk potential) was also provided by the Lipid Research Clinics Hyperlipidemia Prevalence Studies (Fig,. 4) [8]. These disclosed higher average LDL cholesterol levels in premenopausal women taking combination oral contraceptives than in their non-contraceptive taking counterparts but substantially lower LDL cholesterol levels in postmenopausal women taking estrogen replacement alone (overwhelmingly Premarin 1.25 mg/d). With reference to HDL, oral contraceptives appeared to have no net effect upon HDL levels. However, more detailed studies [9] disclosed that though those containing the more highly androgenic progestational components (the 19 non-testosterone derivatives) were associated with lower HDL levels than those containing the less androgenic progestins. Once again postmenopausal estrogen replacement therapy had a substantial effect

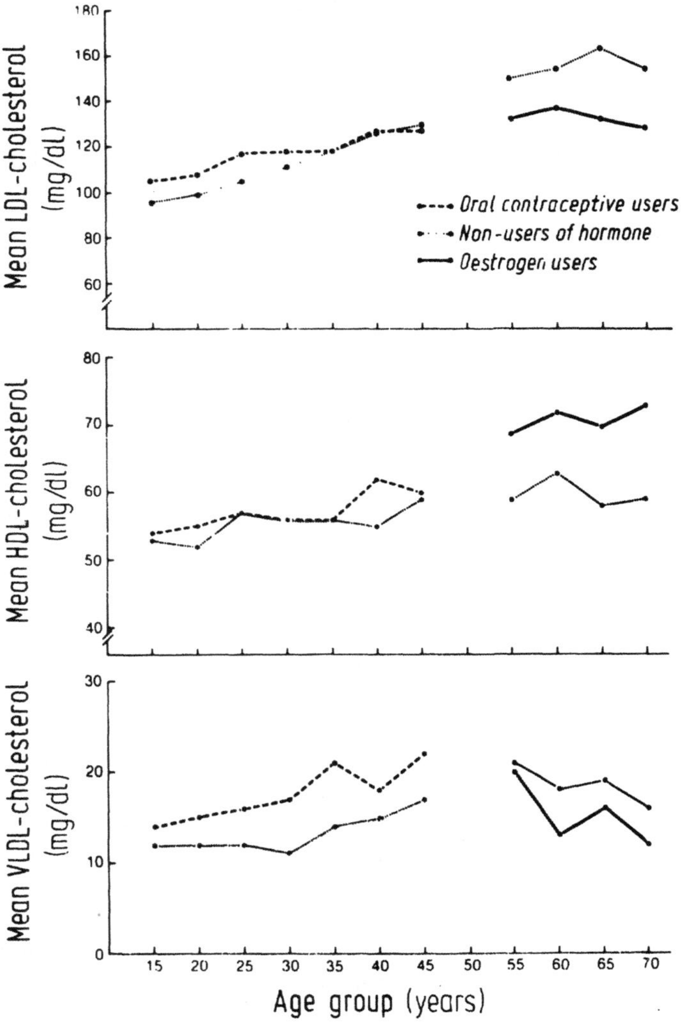

Fig. 4. Plasma lipoprotein cholesterol levels in users and nonusers of oral contraceptives and estrogens. Reprinted with permission, from Wallace RB, et al. (1979) Altered plasma lipid and lipoprotein levels associated with oral contraceptive and oestrogen use. Lancet 2: 114

upon HDL, clearly appearing to raise average levels. Direct tests of these epidemiological associations were provided in a controlled metabolic study [10] in which six clearly estrogen deficient postmenopausal women were fed a diet of constant composition (40% fat, 45% carbohydrates, 15% protein, polyunsaturated/saturated fatty acid ratio 0.85) supplemented with three large egg yolks (total cholesterol intake 1000 mg/d) for 84 days. During the middle 28 of those 84 days ethinyl estradiol 1 mcg/kg/d was added. The major findings of these studies were as follows (Fig. 5) [10]: 1) a rapid rise (by ca. 40%) in plasma triglyceride levels, reflected by triglyceride enrichment of all lipoprotein fractions; 2) a prompt drop in total cholesterol confined to the LDL fraction, average LDL cholesterol levels declining from 173 mg/dl to 130 mg/dl

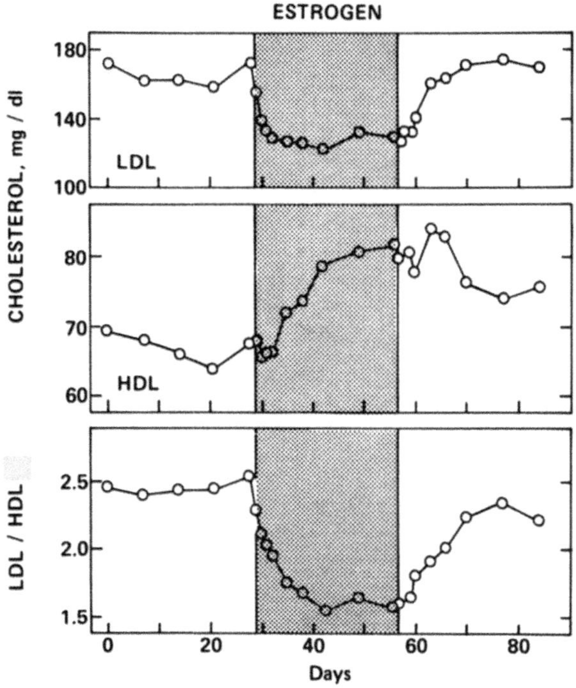

Fig. 5. Mean changes in HDL2 and HDL3 cholesterol levels and postheparin plasma HTGL activity in the six postmenopausal women. The ethinyl estradiol administration (1 ug/kg per day) is indicated by the shaded area from day 28 to 56. Reprinted with permission, from Applebaum-Bowden D, McLean P, Steinmetz A, Fontana D, Matthys C, Warnick GR, Cheung M, Albers JJ, Hazzard WR (1989) Lipoprotein, apolipoprotein, and lipolytic enzyme changes following estrogen administration in postmenopausal women. J Lipid Res 30:1895–1906

within three days and remaining at that level throughout the estrogen supplementation period; 3) a gradual but progressive rise in HDL by an average of ca. 20%, this being confined to the more anti-atherogenic HDL_2 fraction (the LDL/HDL ratio declining by ca. 40% by 28 days on the estrogen). Apolipoprotein levels reflected these changes for the most part: A1 increased substantially and progressively, A2 only marginally, total apo B declined slightly (ca. 10%) and apo E plummeted by 35%. Of potential mechanistic importance, post-heparin hepatic triglyceride lipase (HTGL) activity levels declined rapidly and precipitously, by ca. 70%, remaining low throughout estrogen therapy. The possible relevance of this last finding to the mechanism of the rise in HDL_2 levels was suggested by a strong inverse relationship between HDL_2 cholesterol and post-heparin HTGL activity levels on the final day of estrogen therapy ($r = 0.89$), consistent with the hypothesis that this lipase (which hydrolizes phospholipid as well as triglyceride) may be an important mediator of HDL catabolism.

Similar metabolic studies in postmenopausal women given androgenic anabolic steroids (for postmenopausal osteoporosis) suggest that androgens have metabolic effects opposite those of estrogens. For instance, stanozolol, a 2, 3 nitrizine ring, 17 alpha-methyl anabolic steroid, produced a ca. 50% decline in HDL cholesterol levels during a six week trial, LDL rising concomitantly by ca. 20% (and the average LDL/HDL ratio increasing from a clearly "safe" to a clearly "atherogenic" range [11]). Within HDL the HDL_2 subfraction declined by 85% and HDL_3 by ca. 33%. Simultaneously post-heparin HTGL increased three-fold. In other studies this increase in post-heparin HTGL with stanozolol was found to precede the decline in HDL_2 levels [12], suggesting a mechanistic role for the changes in HTGL in the hypoalphalipoproteinemia induced by anabolic steroids. These contrasting effects were graphically displayed in the single postmenopausal woman sequentially given estrogen and stanozolol individually and together (Fig. 6) [13]: during estrogen HDL increased and LDL decreased; during stanozolol LDL increased and HDL decreased; LDL and HDL levels on both drugs simultaneously were equal to the level off both drugs; and estrogen given in a final phase reproduced the hyperalphalipoproteinemia and hypobetalipoproteinemia seen when she was initially referred while taking estrogen. The role of the induced changes in HTGL in modulating her HDL responses was underscored in these studies: on estrogen, when HTGL declined, HDL catabolism was re-

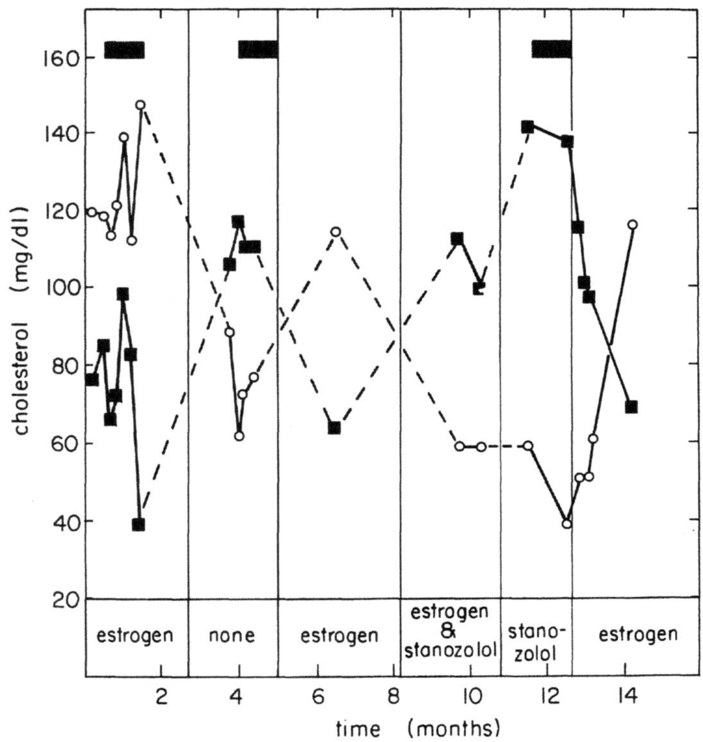

Fig. 6. High-density lipoprotein cholesterol (HDL-C, circles) and low-density lipoprotein cholesterol (LDL-C, squares) levels during treatment with estrogen (ethinyl estradiol, 0.06 mg/day), stanozolol (6 mg/day), both, or neither. Solid bars indicate timing of HDL turnover studies. The subject had been on cyclic estrogen therapy for several years when these studies were initiated. Reprinted with permission, from Hazzard WR, et al. (1984). Preliminary report: Kinetic studies on the modulation of high-density lipoprotein, apolipoprotein, and subfraction metabolism by sex steroids in a postmenopausal woman. Metabolism 33:779–784

tarded; on stanozolol, when HTGL increased, HDL catabolism was accelerated.

Thus in summary exogenous sex steroids appear to influence lipoproteins in the manner that is a caricature of the effects of endogenous sex steroids: estrogens raise HDL (selectively HDL_2) and lower LDL, while androgens lower HDL (selectively HDL_2) and raise LDL; 2) exogenous sex steroids are likely to affect LDL metabolism via effects on the LDL receptor: estrogens increase LDL receptor activity (predominantly shown in animals [14] but recently in humans as well [15]), while androgens may be hypothesized to do the opposite; 3) exogenous sex steroids appear to alter HDL levels predominantly by modulation of HDL catabolism (though A1 synthesis may be also increased [16]): estrogens retard HDL catabolism whereas androgens accelerate HDL catabolism; 4) modulation of HDL (and possibly LDL) metabolism by sex steroids may be mediated by alterations in hepatic triglyceride lipase activity, which appears to facilitate HDL catabolism. All told these changes in lipoprotein metabolism account for a substantial proportion of the sex differential in cardiovascular disease and longevity in Western societies.

References

1. Somers AR (1978) The high cost of health care for the elderly: diagnosis, prognosis, and some suggestions for therapy. Journal of Health Politics, Policy and Law 3:163-180.
2. Hazzard WR (1986) Biological basis of the sex differential in longevity. JAGS 34:455-471.
3. Wylie CM (1984) Contrasts in the health of elderly men and women: an analysis of recent data for whites in the United States. JAGS 32:670-675.
4. Barrett-Connor E, Bush TL (1991) Estrogen and coronary heart disease in women. JAMA 265:1861-67.
5. Holden C (1983) Can smoking explain ultimate gender gap? Science 221: 1034.
6. Sullivan JL (1983) The sex differential in ischemic heart disease. Persp Biol Med 26:657-671.
7. Hazzard WR (1990) A central role of sex hormones in the sex differential in lipoprotein metabolism, atherosclerosis, and longevity. In: Ory MG, Warner HR, ed. Gender, health, and longevity: multidisciplinary perspectives. New York: Springer Publishing Company, 1990:87-108.
8. Wallace RB, Hoover J, Barrett-Connor E, Rifkind BM, Hunninghake DB, Mackenthum A, Heiss G (1979) Altered plasma lipid and lipoprotein levels associated with oral contraceptive and estrogen use. Lancet 2:111-114.
9. Wahl P, Walden C, Knopp R, Hoover J, Wallace R, Heiss G, Rifkind B (1983) Effect of estrogen/progestin potency on lipid/lipoprotein cholesterol. N Eng J Med 308:862-867.
10. Applebaum-Bowden D, McLean P, Steinmetz A, Fontana D, Matthys C, Warnick GR, Cheung M, Albers JJ, Hazzard WR (1989) Lipoprotein, apolipoprotein, and lipolytic enzyme changes following estrogen administration in postmenopausal women. J Lipid Res 30:1895–1906.
11. Taggart HM, Applebaum-Bowden D, Haffner S, Warnick GR, Cheung MC, Albers JJ, Chesnut CH, Hazzard WR (1982) Reduction in high density lipoproteins by anabolic steroid (Stanozolol) therapy for postmenopausal osteoporosis. Metabolism 31:1147–52.

12. Applebaum-Bowden D, Haffner SM, Hazzard WR (1987) The dyslipoproteinemia of anabolic steroid therapy: increase in hepatic triglyceride lipase precedes the decrease in high density $lipoprotein_2$ cholesterol. Metabolism 36:949–952.
13. Hazzard WR, Haffner SM, Kushwaha RS, Applebaum-Bowden D, Foster DM (1984) Preliminary report: kinetic studies on the modulation of high-density lipoprotein, apolipoprotein, and subfraction metabolism by sex steroids in a postmenopausal woman. Metabolism 33:779–784.
14. Windler E, Kovanen Y, Chao S, Brown MS, Hand RJ, Goldstein JL (1980) The estradiol simulated lipoprotein receptor of rat liver: a binding site that mediates the uptake of rat lipoproteins containing apoproteins B & E. J Biol Chem 255: 10464–10471.
15. Ma PT, Yamamoto T, Goldstein JL, Brown MS (1986) Increased mRNA for low density lipoprotein receptor in liver of rabbits treated with 17-alpha ethinyl estradiol. Proc Nat Acad Sci USA 83:792–796.
16. Schaefer EJ, Foster DM, Zech LA, Lingren FT, Brewer HB, Levy RI (1983) The effects of estrogen administration on plasma lipoprotein metabolism in premenopausal females. J Clin Endocrinol Metab 57:262–267.

Induction of the Hepatic Uptake of Lipoproteins by Estrogen

E. WINDLER and S. JÄCKLE

The Effect of Estrogen on the Hepatic LDL-Receptor

Male rats treated with large amounts of 17-α-ethinyl estradiol develop a profound hypolipidemia that affects all of the major lipoprotein classes (Chao et al. 1979). Plasma cholesterol, triglycerides, and the apolipoproteins A-I and E are lowered to less than 10% of normal. The augmented clearance of low density lipoproteins (LDL) is eliminated in functionally eviscerated animals, indicating that the liver is the site of the increased removal (Chao et al.1979). This is in line with the observation that isolated perfused rat livers from estradiol-treated rats catabolize LDL at a much higher rate than controls. Membranes, prepared from livers of estradiol-treated rats, exhibit a 3- to 10-fold increase in saturable binding sites for human LDL (Kovanen et al. 1979). The estradiol-stimulated lipoprotein receptor of the rat recognizes lipoproteins that contain apolipoprotein B-100 (LDL, VLDL, and VLDL remnants) in vivo and in vitro (Windler et al. 1980). The receptor also recognizes synthetic lamellar complexes of lecithin and apolipoprotein E as well as canine high density lipoproteins containing apolipoprotein E (Windler et al. 1980). It is biochemically and immunologically identical with the LDL-receptor of other species (Roach and Noel 1985; Cooper et al. 1987; Jaeckle et al. 1989).

Regulation of the hepatic LDL-receptor is a major factor controlling the plasma concentration of atherogenic lipoprotein species. Depletion of hepatic cholesterol is accompanied by an increased number of LDL-receptors in hepatocytes; conversely, diets rich in cholesterol or saturated fats reduce the number of LDL-receptors and increase plasma cholesterol levels (Brown and Goldstein 1986; Havel and Hamilton 1988). Regulation of the LDL-receptor is achieved in part on the level of gene transcription via sterol-sensitive regulatory elements (Goldstein and Brown 1990), and mRNA for the LDL-receptor is increased by administration of pharmacological doses of estradiol (Ma et al. 1986; Staels et al. 1990). 48 h of feeding an atherogenic diet containing lard, cholesterol, and cholate decreased the number of LDL receptors in estradiol-treated rats by 40%. 2 h after injection of LDL or a constant infusion of β-VLDL, the number of LDL-receptors was reduced by 50%. When β-VLDL were infused into estradiol-treated rats for 17 h, hepatic LDL-receptors decreased even to about 10% (Erickson et al.1989).

The ability of estradiol to induce LDL-receptors was investigated in animals pretreated with an atherogenic diet for 2 weeks; estradiol still increased the number of LDL-receptors fivefold as compared to the number found in livers of untreated animals fed a chow diet; however, the increase was only about 50% of that found in liver membranes from animals on the control diet given estradiol (Erickson et al. 1989). Thus, hepatic LDL-receptors remain inducible by estradiol despite a massive increase in hepatic free and esterified cholesterol produced by feeding the cholesterol-cholate-lard diet. However, the increase in functional LDL-receptors was only half that which would have occurred in estradiol-treated animals on a normal diet. Thus it seems likely that the high concentration of hepatic cholesterol attenuates the induction of receptors by estradiol (Erickson et al. 1989). The observation that estradiol induces enhanced expression of LDL-receptors in cholesterol-loaded livers suggests that estrogens increase the number of LDL-receptors by affecting a different gene regulatory element or that they modify posttranscriptional events in a manner different from sterols (Erickson et al. 1989).

The Effect of Estrogen on the Hepatic Uptake of Apolipoprotein B-Containing Lipoproteins

The proposition that chylomicron remnants are removed by a process distinct from that mediated by the LDL-receptor is based on three observations that need to be explained (Brown and Goldstein 1986; Havel and Hamilton 1988; Mahley 1988): first, the hepatic uptake of chylomicron remnants seems to be normal in humans and in Watanabe-heritable hyperlipidemic (WHHL) rabbits, which express no LDL-receptors or have a severe deficiency of LDL-receptors; second, interventions that affect the number of LDL-receptors have been reported to have little effect on chylomicron remnant removal by the liver; third, antibodies against the hepatic LDL-receptor fail to inhibit binding of chylomicron remnants to liver membranes (Nagata et al. 1988).

However, chylomicron remnants and beta-very low density lipoproteins (β-VLDL) bind to the hepatic LDL-receptor in vitro (Windler et al. 1988; Jaeckle et al. 1989); chylomicron remnant removal is regulated in cultured rat hepatocytes by interventions that change the number of LDL-receptors as inhibitors of 3-hydroxy-3-methylglutaryl coenzyme A (HMG-CoA) reductase, mevalonic acid, insulin, and lipoprotein deficient serum (Jensen et al. 1988 a; Jensen et al. 1988 b; Windler et al. 1988) and excess LDL compete for the uptake of chylomicron remnants by cultured hepatocytes or perfused livers (Windler et al. 1988).

Binding of chylomicron remnants and β-VLDL to hepatic plasma and endosomal membranes is highly induced by estradiol and readily inhibited by excess unlabeled LDL, heparin, and antiserum prepared against the rat or bovine LDL-receptor. LDL, β-VLDL, and chylomicron remnants bind strongly to the LDL-receptor, as demonstrated by ligand blotting (Wade et al. 1986; Windler et al. 1988; Jaeckle et al. 1989). Evidently, chylomicron remnants and VLDL remnants bind at least in vitro to the LDL-receptor.

LDL, β-VLDL, and chylomicron remnants are taken up by the liver of estradiol-treated rats. These particles accumulate first in the compartment of uncoupling of receptor and ligand (CURL) and later in multivesicular bodies (MVB) (Chao et al. 1981, Jones et al. 1984, Jaeckle et al. 1989). The concentration of chylomicron remnants and β-VLDL in isolated endosomes closely resembled that of LDL (Jaeckle et al. 1989), suggesting that estradiol may have also an effect on the uptake of chylomicron remnants and β-VLDL. In normal rats chylomicron remnants accumulate in endosomes at a lower rate than β-VLDL, although chylomicron remnants are rapidly removed from plasma. This suggests that there is a delay in the delivery of chylomicron remnants into the endocytic pathway following initial rapid uptake by the liver of normal rats (Jaeckle et al. 1991a). The reason for this apparent delay in delivery remains to be established. Chylomicron remnants may bind to surface components of the sinusoidal plasma membrane of hepatocytes to be modified before endocytosis occurs. This phenomenon cannot be recognized in estradiol-treated rats. The most obvious difference between estradiol-treated and untreated rats that could account for the differing behavior of chylomicron remnants in the two conditions is the increased number of LDL-receptors in estradiol-treated rats. Therefore we suggest that endocytosis of chylomicron remnants in estradiol-treated rats is mediated predominantly by the LDL-receptor.

To study the contribution of the LDL-receptor to the remnant removal in vivo, turnover studies of chylomicron remnants and β-VLDL were performed in rats after interventions that are known to affect the number of LDL-receptors. Estradiol-treatment stimulates the plasma removal of chylomicron remnants and β-VLDL substantially, when saturating dosages are injected. Using only tracer amounts, as has been done in previous publications, the stimulation of the hepatic chylomicron remnant and β-VLDL removal becomes undetectable, because hemodynamic parameters and receptor affinity, which is not altered by estradiol-treatment, become rate limiting, and the increased number of LDL-receptors fails to further stimulate the already rapid remnant removal. The receptor-mediated endocytosis of asialofetuin, a ligand of the asialoglycoprotein receptor, is unaltered by estradiol-treatment, indicating that the stimulation of the lipoprotein uptake is not due to a nonspecific induction of the hepatic endocytosis. The removal of chylomicron remnants and β-VLDL is significantly delayed in rats fed an atherogenic diet containing lard, cholesterol and bile acids. Liver perfusion experiments indicate that the reduced remnant removal in these rats is primarily caused by down-regulation of hepatic LDL receptors. This finding further supports the concept that the LDL-receptor plays an important role not only for LDL, but also for the rate of uptake of remnants of triglyceride-rich lipoproteins under physiological conditions.

The LDL-receptor related protein (LRP), which was shown to have sequence identity with the a_2-macroglobulin-receptor (Kristensen et al. 1990; Strickland et al. 1990), was proposed to mediate the uptake of apolipoprotein E-containing lipoproteins (Herz et al. 1988; Beisiegel et al. 1989; Kowal et al.

1989; Lund et al. 1989). Unlike the LDL-receptor, this protein is not regulated by estradiol or cholesterol (Kowal et al. 1989; Lund et al. 1989).

Asialofetuin, asialomucin, and an antibody against the asialoglycoprotein receptor were shown to inhibit binding, internalization, and degradation of chylomicron remnants in the human hepatoma cell line HepG2 up to 70% (Windler et al. 1991). However, in the rat, plasma removal of chylomicron remnants was not inhibited by asialofetuin and vice versa, uptake of asialofetuin was not inhibited by chylomicron remnants or β-VLDL in vivo. Both proteins, LRP and the asialoglycoprotein receptor, which were shown to bind apolipoprotein E-containing lipoproteins in vitro, are not regulated by estradiol or an atherogenic diet in vivo. Their contribution to the removal of lipoproteins in vivo remains to be established. Thus, the effect of interventions, which regulate the number of LDL-receptors, on chylomicron remnant turnover, points to an important function of the LDL-receptor for chylomicron remnant removal in vivo.

The Effect of Estrogen on the Hepatic Uptake of HDL

Apolipoprotein E-depleted HDL-particles are internalized by rat livers, transported to early (CURL) and late (MVB) endosomes and are finally subjected to lysosomal degradation. The receptor recycling endosomal compartment (RRC), whose membranes contain a high concentration of recycling receptors and transferrin (Jaeckle et al. 1991b), which is thought to recycle to the cell surface bound to its receptor, failed to accumulate HDL-particles after injection of HDL. HDL-particles accumulate, however, in multivesicular bodies (MVB), the immediate prelysosomal compartment and in lysosomes at the same extent as ligands, which are known to be degraded within lysosomes. Thus, for intact HDL particles we found internalization into endosomes, yet no evidence for retroendocytosis.

HDL-associated cholesteryl esters are taken up in excess of the uptake of HDL-associated apolipoprotein A-I indicating selective uptake of HDL-cholesteryl esters. Cholesteryl esters selectively taken up from HDL are internalized into hepatocytes via a non-endosomal, non-lysosomal pathway, and do not accumulate in CURL, MVB, or lysosomes. Cholesteryl esters accumulate, however, in a cytosolic fraction, whose major proteins are demonstrated to differ from those derived from endosomes, the Golgi apparatus, lysosomes and plasma membranes.

Estradiol-treatment had no significant effect on the liver uptake of apolipoprotein A-I of apolipoprotein E-deficient HDL_3 indicating that the LDL-receptor does not contribute to the hepatic uptake of HDL-particles. However, the selective uptake of cholesteryl esters was substantially higher in estradiol-treated rats. Yet, in vitro selective binding of cholesteryl esters from HDL to rat liver membranes is not stimulated by previous estradiol-treatment. Therefore the enhancement of the selective uptake in vivo seems to be independent of the LDL-receptor and the membrane composition and may be regulated by a changed hepatic cholesterol content and compartimentalization of cholesterol after estradiol-treatment. Thus, particle-uptake of HDL via receptor mediated endocytosis and selective uptake of cholesteryl esters from HDL are regulated independently, as suggested by previous findings (Rinninger and Pittman 1987; Rinninger and Pittman 1988; Rinninger and Greten 1990).

Apolipoprotein E-containing HDL-particles may also bind to the LDL-receptor. Indeed, the hepatic uptake of HDL particles was shown to be regulated by estradiol according to their content in apolipoprotein E (Chao et al. 1979), supporting that their removal is mediated by the LDL-receptor if the content in apolipoprotein E is sufficient. The rate of metabolism of labelled apolipoprotein E incorporated in liposomes was inducible by estradiol, but not that of apolipoprotein A-I in liposomes. This is in line with the observation that the ability to induce the catabolism of HDL depends on their contents in apolipoprotein E. Thus, the appreciable decrease of apolipoprotein A-I in the plasma of estradiol-treated rats apparently is a consequence of the induced catabolism of HDL-particles containing both, apolipoprotein E and A-I. Despite their high contents in apolipoprotein E, rat HDL_C is taken up by rat livers at a much slower rate than β-VLDL or chylomicron remnants. Nonetheless, their uptake is still regulated by estradiol to a similar extent as LDL, β-VLDL, or chylomicron remnants.

Summary

In conclusion, all lipoproteins that contain either apolipoprotein B-100 or E or both can be taken up by the liver via the LDL-receptor. Estradiol-treatment increases the number of LDL-receptors and thus may influence the rate of removal of LDL, remnants of triglyceride-rich lipoproteins and apolipoprotein E-containing HDL. Clinically this may be the mechanism underlying the higher cholesterol concentrations in states of estrogen deficiency and the subsequent more rapid progression of arteriosclerosis.

References

Beisiegel U, Weber W, Ihrke G, Herz J and Stanley KK (1989) The LDL-receptor-related protein, LRP, is an apolipoprotein E-binding protein. Nature 341:162–164.

Brown MS, Goldstein JL (1986)A receptor-mediated pathway for cholesterol homeostasis. Science 232:34–47.

Chao Y-s, Windler EE, Chen GC, Havel RJ (1979) Hepatic catabolism of rat and human lipoproteins in rats treated with 17 a-ethinyl estradiol. J Biol Chem 254: 11360–11366.

ChaoY-s, Jones AL, Hradek GT, Windler ETT, Havel RJ (1981) Autoradiographic localization of the sites of uptake, cellular transport and catabolism of low density lipoproteins in the liver of normal and estrogen-treated rats. Proc Natl Acad Sci USA 78:597–601.

Cooper AD, Nutik R, Chen J (1987) Characterization of the estrogen-induced lipoprotein receptor of rat liver. J Lipid Res 28:59–68.

Erickson SK, Jaeckle S, Lear SR, Brady SM, Havel RJ (1989) Regulation of hepatic cholesterol and lipoprotein metabolism in ethinyl estradiol-treated rats. J Lipid Res 30:1763–1771.

Goldstein JL, Brown MS (1990) Regulation of the mevalonate pathway. Nature 343: 425–430.

Havel RJ, Hamilton RL (1988). Hepatocytic lipoprotein receptors and intracellular lipoprotein catabolism. Hepatology 8:1689–1704.

Herz J, Hamann U, Rogne S, Myklebost O, Gausepohl H, Stanley KK (1988) Surface location and high affinity for calcium of a 500-kd liver membrane protein closely related to the LDL-receptor suggest a physiobgical role as lipoprotein receptor. 7: 4119–4127.

Jaeckle S, Brady SE, Havel RJ (1989) Membrane binding sites for plasma lipoproteins on endosomes from rat liver. Proc Natl Acad Sci USA 86:1880–1884.

Jaeckle S, Runquist E, Brady S, Hamilton RL, Havel RJ (1991a) Isolation and characterization of three endosomal fractions from the liver of normal rats after lipoprotein loading. JLipid Res, in press.

Jaeckle S, Runquist EA, Miranda-Brady S, Havel RJ (1991b) Trafficking of the epitermal growth factor receptor and transferrin in three hepatocytic endosomal fractions. JBiolChem 266:1396–1402.

Jensen E, Floren C-H, Nilsson A (1988a) Insulin stimulates the uptake of chylomicron remnants in cultured rat hepatocytes. Eur J Clin lnvest 18:226–232.

Jensen E, Floren C-H, Nilsson A (1988b) Lipoprotein-deficient serum stimulates the uptake of chylomicron remnants in cultured rat hepatocytes. Scand J Clin Lab Invest 48:553–556.

Jones AL, Hradek GT, Hornick C, Renaud G, Windler EET, Havel RJ (1984) Uptake and processing of remnants of chybmicrons and very low density lipoproteins by rat liver. J Lipid Res 25:1151–1158.

Kovanen PT, Brown MS, Goldstein JL (1979) Increased binding of low density lipoprotein to liver membranes from rats treated with 17 a-ethinyl estradiol. J Biol Chem 254:11367–11373.

Kowal RC, Herz J, Goldstein JL, Esser V, Brown MS (1989) Low density lipoprotein receptor-related protein mediates uptake of cholesteryl esters derived from apoprotein E-enriched lipoproteins. Proc Natl Acad Sci USA 86:5810–5814.

Kristensen T, Moestrup SK, Gliemann J, Bendtsen L, Sand O, Sottrup-Jensen L (1990) Evidence that the newly cloned bw-density-lipoprotein receptor related protein (LRP) is the a_2-macroglobulin receptor. FEBS-Lett 276:151–155.

Lund H, Takahashi K, Hamilton RL, Havel RJ (1989) Lipoprotein binding and endosomal itinerary of the low density lipoprotein receptor-related protein in rat liver. Proc Natl Acad Sci USA 86:9318–9322.

Ma PTS, Yamamoto T, Goldstein JL, Brown MS (1986) Increased mRNA for low density lipoprotein receptor in livers of rabbits treated with 17 a-ethinyl estradiol. Proc Natl Acad Sci USA 83:792–796.

Mahley RW (1988). Apolipoprotein E: cholesterol transport protein with expanding role in cell bioiogy. Science 240:622–630.

Nagata Y, Chen J, Cooper AD (1988) Role of low density lipoprotein receptor dependent and -independent sites in binding and uptake of chylomicron remnants in rat liver. J Biol Chem 263:15151–15158.

Rinninger F, Greten H (1990) High density lipoprotein particle uptake and selective uptake of high density lipoprotein-associated cholesteryl esters by J774 macrophages. Biochim Biophys Acta 1043:318–326.

Rinninger F, Pittman RC (1987) Regulation of the selective uptake of high density lipoprotein-associated cholesteryl esters. J Lipid Res 28:1313–1325.

Rinninger F, Pittman TC (1988) Regulation of the selective uptake of high density lipoprotein-associated cholesteryl esters by human fibroblasts and Hep G2 hepatoma cells. J Lipid Res 29:1179–1194.

Roach PD, Noel S-P (1985) Solubilization of the 17 a-ethinyl estradiol-stimulated low density lipoprotein receptor of male rat liver. J Lipid Res 26:713–720.
Staels B, Jansen H, Van Tol A, Stahnke G, Will H, Verhoeven G, Auwerx J (1990) Development, food intake, and ethinylestradiol influence hepatic triglyceride lipase and LDL receptor mRNA levels in rats. J LipW Res 31:1211–1218.
Strickland DK, Ashcom AD, Williams S, Burgess WH, Migliorini M, Argraves WS (1990) Sequence identity between the a_2-macroglobulin receptor and low density lipoprotein receptor-related protein suggests that this molecule is a multifunctional receptor. J Biol Chem 265:17401–17404.
Wade DP, Knight BL, Soutar AK (1986) Binding of low-density lipoprotein and chylomicron remnants to the hepatic low-density lipoprotein receptor of dogs,rats and rabbits demonstrated by ligand blotting. Failure to detect a distinct chylomicron-remnant-binding protein by ligand blotting. Eur J Biochem 159:333–340.
Windler EET, Greeve J, Daerr WH, Greten H (1988) Binding of rat chylomicrons and their remnants to the hepatic low-density-lipoprotein receptor and its role in remnant removal. J Biol Chem 252:553–561.
Windler E,Greeve J, Levkau B, Kolb-Bachofen V, Daerr W, Greten H (1990) The human asialoglycoprotein-receptor: a possible binding site for LDL and chylomicron remnants. Biochem J in press
Windler EET, Kovanen PT, Chao Y-s, Brown MS, Havel RJ, Goldstein JL (1980) The estradiol-stimulated lipoprotein receptor of rat liver. A binding site that mediates the uptake of rat lipoproteins containing apoprotein B and E. J Biol Chem 255: 10464–10471.

Changes in Serum Lipoprotein Profiles During Puberty – A Longitudinal Study of Dresden Schoolchildren

W. JAROSS, A. TRÜBSBACH, and M. BORRMANN

The typical sex-difference in the lipoprotein profile between females and males aging 20–45 years is characterized by higher TC (*), LDL-C, and Apo B, lower HDL-C, Apo AI, HDL-2C and HDL-3C in males, whereas no statistically proved differences in regard to serum concentrations of Apo AII, D, E could be observed in our and other studies [1, 5]. It is commonly supposed that these differences are the result of sexual maturation. With respect to the sex-difference of HDL variables it could be demonstrated in several studies that in the course of sexual maturation the HDL-C level decreases in boys, whereas no or only small changes could be found in girls [4, 11, 12]. On the other hand the TC, LDL-C and Apo B serum concentrations show very untypical sex-differences immediately after puberty. In contrast to the common assumption these variables are higher in females than in males between the age of 16 and the beginning of the twenties [2, 8, 9]. In both sexes these variables increase after puberty. Because the sexual maturation in girls is finished 1–2 years earlier than in males the unexpected sex-difference could be the result of that age difference in sexual maturation at least in part.

There are also other questions of interest: Which changes of main apolipoprotein concentrations can be observed in the course of puberty? Which correlations exist between lipids, Apos and sex-hormones? From the point of view of CAD prevention an extremely interesting aspect is the tracking of high or low lipid and Apo concentrations during puberty, and only little information is available especially for the Apos. To answer some of these questions we carried out a longitudinal study of Dresden schoolchildren over 6 years where we analysed serum lipid and Apo concentrations every year.

Material and Methods

223 children from 3 primary schools in the town of Dresden were investigated yearly in the summers from 1981 to 1987. At the beginning of the study the pupils were between 9 and 13 years old. The investigation was finished with the end of the school time when the pupils were 16 or 17 years old.

(*) *Abbreviations:* TC = total cholesterol; LDL-C = low density lipoprotein cholesterol; HDL-C = high density lipoprotein cholesterol; Apo AI, AII, B, D, E = apolipoprotein AI, AII, B, D, E; CAD = Coronary artery disease

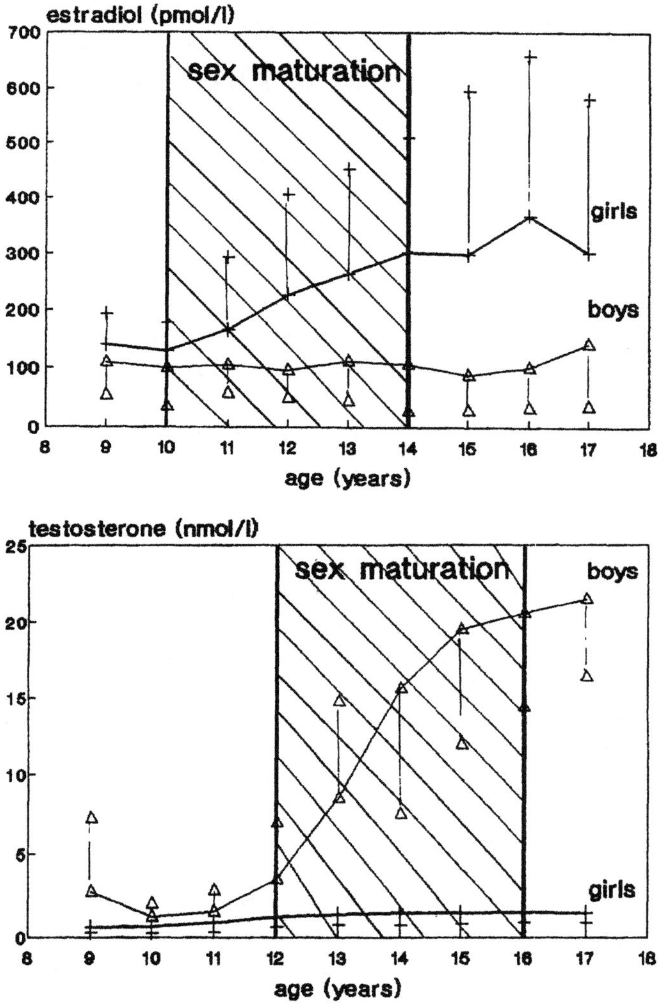

Fig. 1. Changes of testosterone and estradiol serum concentrations during puberty

The children were investigated by a pediatrician every year and the stage of sexual maturation was estimated according to the Tanner staging.

The following variables were analyzed in blood serum after overnight fast: TC, TG, HDL-C, LDL-C (calculated according to the FRIEDEWALD-formula) Apo AI, AII, B, testosterone and estradiol. Some anthropometric data (height, weight) were measured. Serum was kept frozen (–20 °C) before analyses, and a strict quality control system was used. (Details of the methods in [10]).

Results

From the 11th to 14th year of age the difference in sexual development between females and males was one Tanner-degree on average. The testosterone serum level in boys increased significantly at the age of 12 years and reached concentrations typical of adults at an age of 16 years on average, whereas the estradiol serum concentrations in girls increased at an age of 10 years and reached the adult level at 14 years (Fig. 1).

TC and LDL-C decreased significantly in boys during puberty, whereas there was only a small change in girls (Fig. 2). The difference became significant toward the end of puberty and girls had significantly higher values also after adjusting for sexual maturation. TC was elevated to about 0,7 mmol/l and LDL-C to about 0,6 mmol/l in girls at the end of sexual maturation.

Apo B serum concentrations increased in both sexes toward the end of puberty without any statitically proved difference between females and males (Fig. 3).

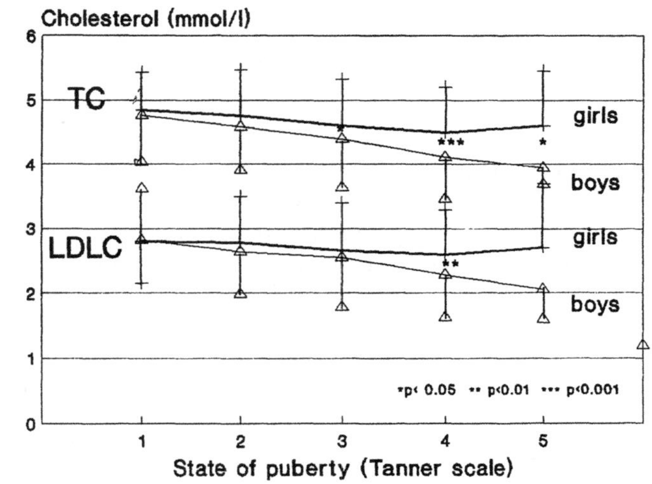

Fig. 2.
Changes of TC and LDL-C serum concentrations during puberty

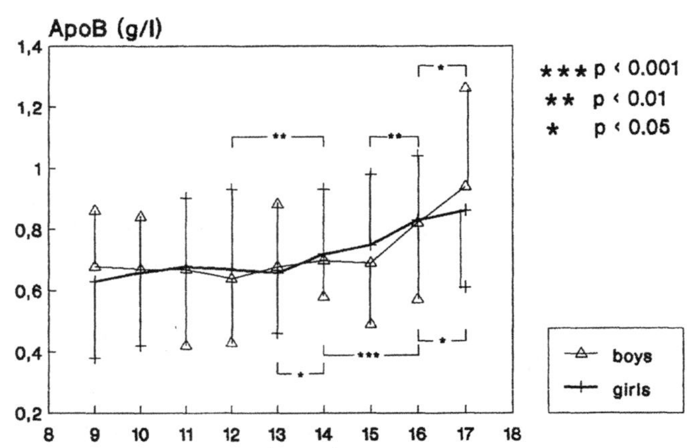

Fig. 3.
Changes of Apo B serum concentrations during puberty

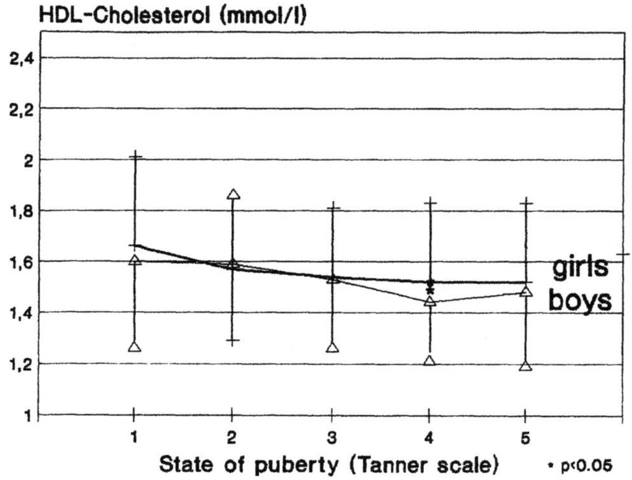

Fig. 4. Changes of HDL-C serum concentrations during puberty

A decrease of HDL-C could be found in both genders, but it was stronger in boys toward the end of puberty and the sex-difference became significant (Fig. 4). The course of Apo AI and AII-serum levels indicate a more pronounced increase of Apo AI in females and that of Apo AII in males without statistical significance (not demonstrated), and as a consequence of these changes the ratio Apo AI/AII showed the typical sex-difference at the end of pube at higher values in females (Fig. 5).

The TG changes during puberty did not lead to statistically significant sex-differences. The linear regression analysis indicates that testosterone might have a stronger influence on some lipoprotein variables than estradiol (Table 1).

The tracking of low or high lipoprotein variables were estimated by calculation of the percentage of values which were found in the same quintile at the end of the study as analysed at the beginning (4 years period on average). The tracking during the period of sexual maturation was 54 to 71% for Apo B, LDL-C TC, HDL-C and Apo AI/AII, whereas it was much smaller for TG and Apo AI (Table 2).

Discussion

In the course of sexual maturation TC and LDL-C decrease in both boys and girls, but much stronger in boys [9]. The higher TC and LDL-C serum concentrations in female adolescents are also existent after adjusting to the state of puberty. But Apo B serum concentrations were not significantly different between both sexes at the end of puberty. That means the typical sex-differences in adults with higher values of Apo B, LDL-C and TC in males are not directly the result of sexual maturation. They develop later at the beginning of the twenties, and the typical male and female hormonal state might not be the dominant cause of these differences. On the other hand, the sex-differ-

Table 1. Correlation between sex-hormones and serum lipid and Apo concentration

		r	
		male	female
E_2	– LDL-C	–0.040	0.094[a]
	– B	0.024	0.041
	– HDL-C	–0.046	–0.032
	– AI	0.042	0.093[a]
	– AII	–0.058	0.047
	– AI/AII	0.202[c]	0.058
	– TG	–0.044	–0 097[a]
	– 6MI	0.043	0.222[c]
T	– TC	–0.241[c]	0.035
	– B	0.023	0.131[b]
	– HDL-C	–0.192[c]	0.083
	– AI	0.201[c]	0.105
	– AII	0.276	0.071
	– TG	0.079	–0 037
	– BMI	0.444[c]	0.192[c]

[a] p<0.05; [b] p<0-.01; [c] p<0.001

ences in HDL variables develop during sexual maturation by a stronger decrease of HDL-C in boys and a tendency for stronger increase in Apo AI in girls and of Apo AII in boys resulting in sex-dependent ratios for Apo AI/AII at the end of puberty.

The decrease of HDL-C in males during puberty is reported by several investigators [4, 9, 11]. In a cross sectional study [4] we reported that the sex-difference in the Apo AI/AII ratio which is linked to the typically different HDL-2/HDL-3 ratio in males and females developed also during puberty, and in the present study we can confirm that result also after adjusting to the state of puberty. In agreement with Kirkland et al. [6] we concluded that testo-

Table 2. Tracking of high or low lipid and Apo serum conc. during puberty

	Values in upper or lower quintile at age 14-17 in % of upper or lower quintile at age 9-12	
	male (%)	female (%)
TC	71	71
LDL-C	67	68
Apo B	62	54
TG	36	48
HDL-C	69	55
Apo AI	33	26
AI/AII	62	69

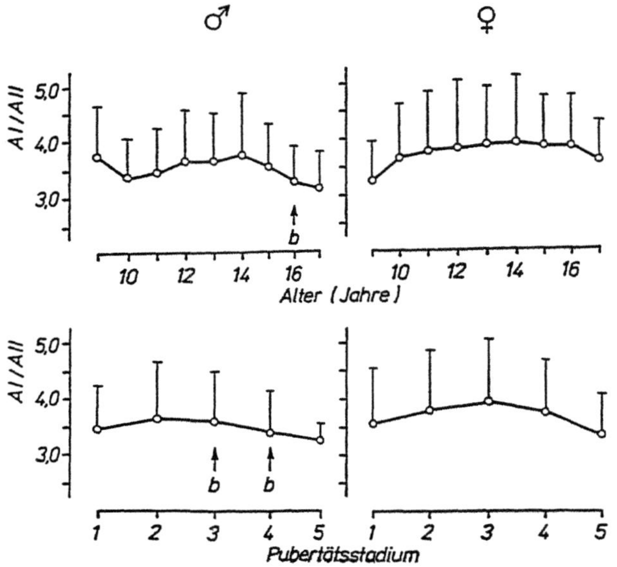

Fig. 5. Changes of Apo AI/AII ratio during puberty
b: $p<0.01$ boys vs. girls

sterone should be considered an important determinant of the sex-typical changes of HDL variables during sexual maturation.

The tracking of lipoprotein variables during childhood and adult life has been described in several studies [3, 7] because the predictive value of childhood measurements regarding the adult coronary risk level is of pronounced interest. Especially in respect to tracking through the puberty clear information is desirable because of the changes in the lipoprotein metabolism in that period. In the Muscatine study [7] 43% children with cholesterol levels greater than the 90th percentile in a single measurement were found to have levels greater than the 90th percentile and 62% greater than the 75th percentile at 20 to 30 years. In contrast to that HDL-C determinations in childhood did not have any predictive value. We found comparable results for TC and LDL-C, but in contradiction to the above-mentioned investigators also a significant tracking of HDL-C. But whereas we investigated only the tracking through the sexual maturation period itself, the period was much longer in the Muscatine study. One possible explanation for the different results is that factors having a very strong influence on the HDL serum concentrations like contraceptives, alcohol consumption and smoking grow in importance between the end of puberty and the beginning of the twenties.

Summary

Lipoprotein variables (TC, TG, HDL-C, LDL-C (calculated), Apo AI, AII and B) sex hormones (T, E_2) and the stage of sexual maturation according to the Tanner scale were determined in a longitudinal study of 223 Dresden schoolchildren from 1981 to 1987. In the course of sexual maturation TC and LDL-C decreased, but in boys much stronger than in girls. Also after adjusting the stage of puberty these sex-untypical differences are significant, indicating that the developing sex-typical hormonal state is not the dominant cause of the differences in the TC and LDL-C serum levels between males and females in later life. On the other hand, the HDL-C as well as the Apo AI/AII ratio decreases in males, indicating that the typical male-female differences are the result of sexual development. According to the linear regression analyses the influence of testosterone on the lipoprotein variables appears to be more remarkable than that of estradiol. The tracking of high serum concentrations of TC, LDL-C and Apo B as well as low values of the HDL-C and Apo AI/AII ratios during puberty was 54 to 71%, indicating a remarkable predictive value of the measurements of those variables in childhood.

References

1. Beaglehole R,Trost DC,Tamir I,Kwiterovich P, Glueck CJ, Insull W, Christensen B (1980) Plasma high-density lipoprotein cholesterol in children and young adults. The Lipid Research Clinics Program. Circulation 62 (Suppl IV):IV–83, IV–92
2. Berenson GS, Srinivasan SR, Hunter SMD, Nicklas TA, Freedman DS, Shear CL, Webber LS (1989) Risk factors in the early life as predictor of adult heart disease.The Bogalusa Heart Study. Am J Med Sci 141–151
3. Clarke WR, Schrott HG, Leaverton PE, Connor WE, Lauer RM (1978) Tracking of blood lipids and blood pressure in school age children: The Muscatine Study. Circulation 58:626–634
4. Jaross W, Baehrecke M, Seiler E (1981) Effects of sexual maturation of serum lipoproteins. Endokrinologie 78:28–34
5. Jaross W, Schulte H, Bergmann S, Assmann G (1991) Differences in CAD risk profile between "West" and "East"-Germany. In: Advances in Lipoprotein and Atherosclerosis Research, Diagnostics and Treatment: Proceedings of the 7th International Dresden Lipid Symposium. Fischer, Jena, p. 171–179
6. Kirkland RT, Keenan BS, Probstfield JL, Patsch W, Lin T-L, Clayton GW, Insull W. Decrease in plasma high-density lipoprotein cholesterol levels at puberty in boys with delayed adolescence: Correlation with testosterone levels. J Am Med Ass 257:502–507
7. Lauer RM, Le J, Clarke WR (1988) Factors affecting the relationship between childhood and adult cholesterol levels: The Muscatine Study. Pediatrics 82:309–18
8. NIH, Publication (1980) 80–1527, p. 90
9. Resnicow K, Morley-Kotchen J, Wynder E (1989) Plasma cholesterol levels of 6585 children in United States: Results of the Know Your Body Screening in five states. Pediatrics 84:969–976
10. Trübsbach A, Borrmann M, Jaross W, Schollberg K, Koch R (1989) Einfluß der Pubertät auf Serumlipide und -apolipoproteine bei 9- bis 17jährigen Schulkindern. Z klin Med 44:327–332
11. Viikari J, Rönnemaa T, Seppänen A, Marniemi J, Porkka K, Räsänen L, Uhari M, Salo MK, Kaprio EEA, Nuutinen EM, Pesonen E, Pietikäinen M, Dahl M, Akerblom HK (1991) Ann Med 23:53–59
12. Williams CL, Carter BJ, Wynder E (1981) Prevalence of selected cardiovascular and cancer risk factors in a pediatric population: The "Know Your Body" project, New York. Prev Med 10:235–250

Coronary Risk in Women: Potential Role of Age, Lipoproteins, Haemostasis, and Hormones

A. von Eckardstein, M. Sandkamp, A. Dirkes-Kersting,
J. Heinrich, D. Bänsch, H. Schulte, and G. Assmann

Introduction

In industrialized countries, cardiovascular diseases are the leading cause of death in both men and women. However, men mostly suffer from myocardial infarction (MI) in the fifth and sixth decades of life, whereas during this period incidence of MI is low in women but increases dramatically after age 55. The reason for these sex-specific differences in the development of coronary heart disease (CHD) is currently unknown but has mostly been adressed to the protective effect of estrogens in women because coronary risk increases severely after both physiological menopause and ovarectomy (reviewed in Bush et al. 1988, Corrao et al. 1990). However, to date the antiatherogenic mechanism of estrogens is poorly understood. To explore possible mechanisms, we analysed lipids, hemostatic, and endocrine parameters in female and male participants of the Prospective Cardiovascular Munster (PROCAM) study. The following questions were asked: 1)Which risk parameters change differently in men and women throughout age? 2) Are there sex-specific statistical interrelationships between these parameters? 3) Do non-sex hormones affect some risk parameters sex-specifically?

Lipids

After age 45, total cholesterol serum concentrations severely increase in women and reach higher mean values than in men (Fig. 1) (Assmann 1989). Results from the Framingham Study suggest that in women the hypercholesterolemia associated risk is not as high as in men. Cholesterol concentrations below 265 mg/dl did not vary between women with and without CHD (reviewed in Bush et al. 1988). Therefore it is a matter of debate whether or not increasing cholesterol concentrations per se mediate the increased coronary risk of postmenopausal women. By contrast to total cholesterol concentration, HDL-cholesterol concentration was a strong, negative, and independent coronary risk factor in women participating in the Framingham-, the LRC-Follow-Up- and the Donolo-Tel Aviv-Studies (reviewed in Bush et al. 1988). However, none of these studies nor the PROCAM study found HDL-cholesterol concentrations decreasing in aging women (Fig. 1). Also menopause does not discriminate between high and low HDL-cholesterol concentrati-

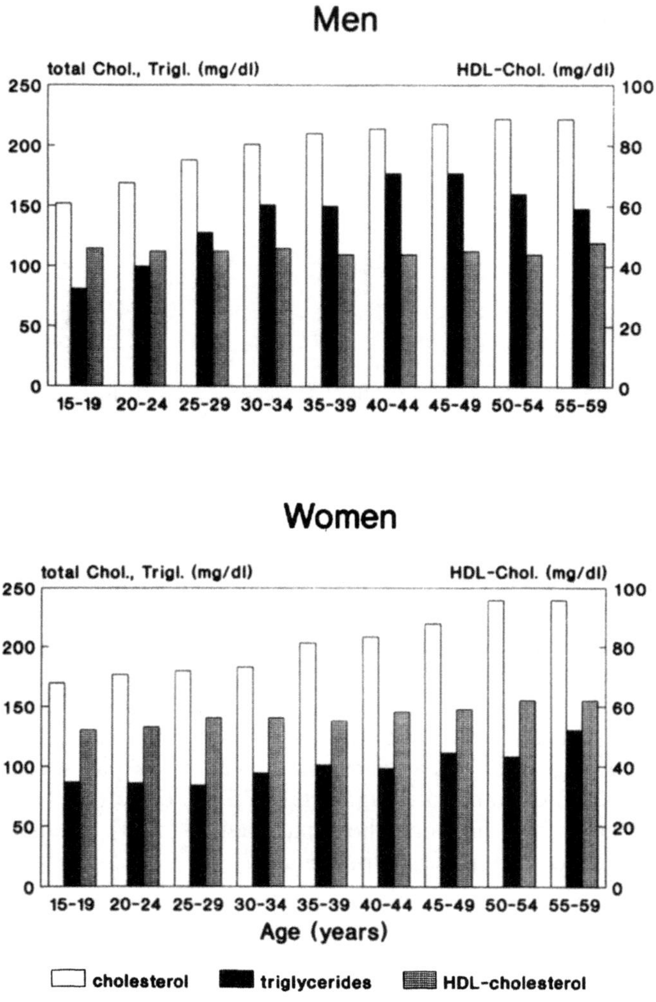

Fig. 1. Age related changes in the serum concentrations of total cholesterol, triglycerides and HDL-cholesterol in male and female participants of the PROCAM study. Note that cholesterol concentrations increase at different life periods in men and women and that after age 50 cholesterol concentrations in women exceed those in men. HDL cholesterol does not adversely change throughout age in either sex

ons. This is in contrast to the clinical finding that HDL-cholesterol concentrations are increased by postmenopausal hormone replacement, a contradicition that is not solved to date (reviewed in Bush et al. 1988, Corrao et al. 1990).

In the PROCAM population significant differences between men and women exist for the age related variation of LDL-cholesterol and Lp(a) serum concentrations: In men, Lp(a) concentrations remain constant throughout life, and LDL-cholesterol concentrations increase until age 45 and then remain constant. In women, however, Lp(a) and LDL-cholesterol concentrations are severely increasing after age 45 and then even exceed the concentrations found in men (Fig. 2). Also in contrast to men, Lp(a) serum concentrations are positively correlated to the body mass index in women (Table 1) (Heinrich et al. 1991a).

Lp(a) and Hemostasis

In several case control studies, Lp(a) concentrations have been positively related to coronary risk in men. To date, the underlying pathomechanism is not understood. Because of the structural homology between the protein compo-

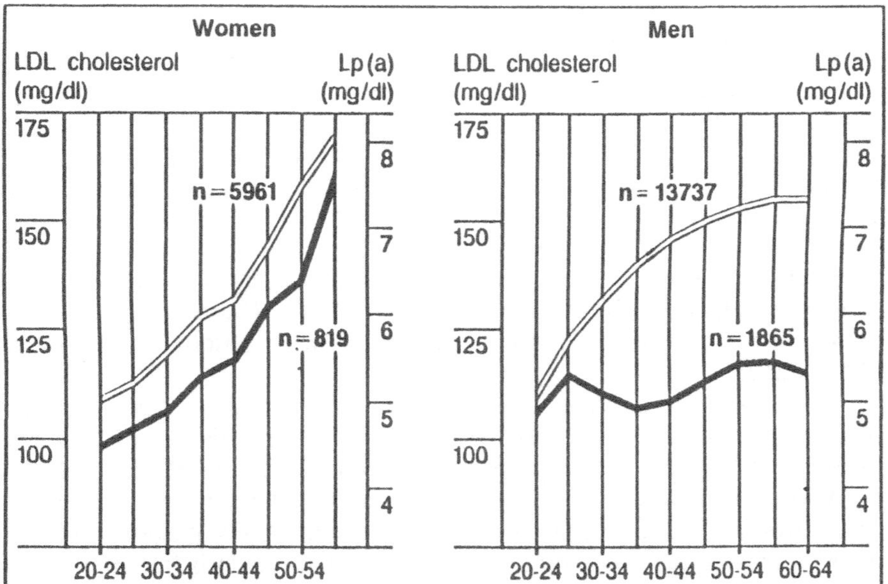

	Women: age 30–44 (mg/dl)	Women: age 45–59 (mg/dl)	Men: age 30–44 (mg/dl)	Men: age 45–64 (mg/dl)
Δ LDL (median)	12	37	14	7
Δ Lp(a) (median)	0.6	2.1	−0.09	0.43

Fig. 2. Age related changes in the serum concentrations of LDL-cholesterol (bright line) and Lp(a) (dark line) in women and men from the PROCAM population. Note the markedly increasing concentrations of both LDL-cholesterol and Lp(a) in women after age 45 which do not occur in men

Table 1. Relationships between in Lp(a) and haemostatic and other metabolic variables as well as established risk factors of coronary heart disease by means of bivariate regression analysis
(r = Pearson correlation coefficient; * $p<0.05$; ** $p<0.01$; *** $p<0.001$)

		r	
		male	female
ln Lp(a)	– cholesterol	0.13***	0.24***
	– systolic b.p.	0.07*	0.15**
	– age	0.05	0.23***
	– body mass index	0.01	0.19***
	– triglycerides	0.04	0.08
	– HDL cholesterol	−0.03	−0.01
	– smoking	0.01	−0.02
	– glucose	−0.01	−0.02
	– fibrinogen	0.08*	0.20***
	– factor VIIc	0.01	0.07
	– protein C	0.05	0.17**
	– AT III	0.01	0.15**
	– EFA	−0.06	−0.03
	– t-PA	−0.07	0.05
	– PAI-1	−0.06	−0.11
	– d-dimer	0.05	0.03

nent in Lp(a) – apo(a) – and plasminogen, it has been suggested that Lp(a) interferes with hemostasis (reviewed in Loscalzo 1990). Therefore, we analysed the relationship between Lp(a) concentrations and hemostatic variables (Tables 1 and 2).

Upon bivariate analysis, Lp(a) and fibrinogen concentrations correlate positively in women and less in men (Table 1). The positive correlation was assessed both in pre- and postmenopausal women. By multivariate analysis Lp(a) concentrations negatively correlated with PAI-1 activity, both in men and in women (Table 2). Since the plasma concentrations of Lp(a) and fibrinogen increased in women older than 50 (Fig. 3), these statistical interdependences are very interesting in regard to coronary risk in women: In the pros-

Table 2. Regression coefficients and amount of the variance explained (in brackets) of ln Lp(a) related to cholesterol, PAI-1 and sys tolic blood pressure (variables taken into account were all parameters listed in table 2; * $p<0.05$, ** $p<0.01$,*** $p<0.001$)

		r	
		male	female
ln Lp(a)	– cholesterol	0.18*** (2.3%)	0.17** (5.8%)
	– PAI-1	−0.12** (1.4%)	−0.14* (2.0%)
	– systolic blood pressure	0.08* (1.1%)	0.11* (1.7%)

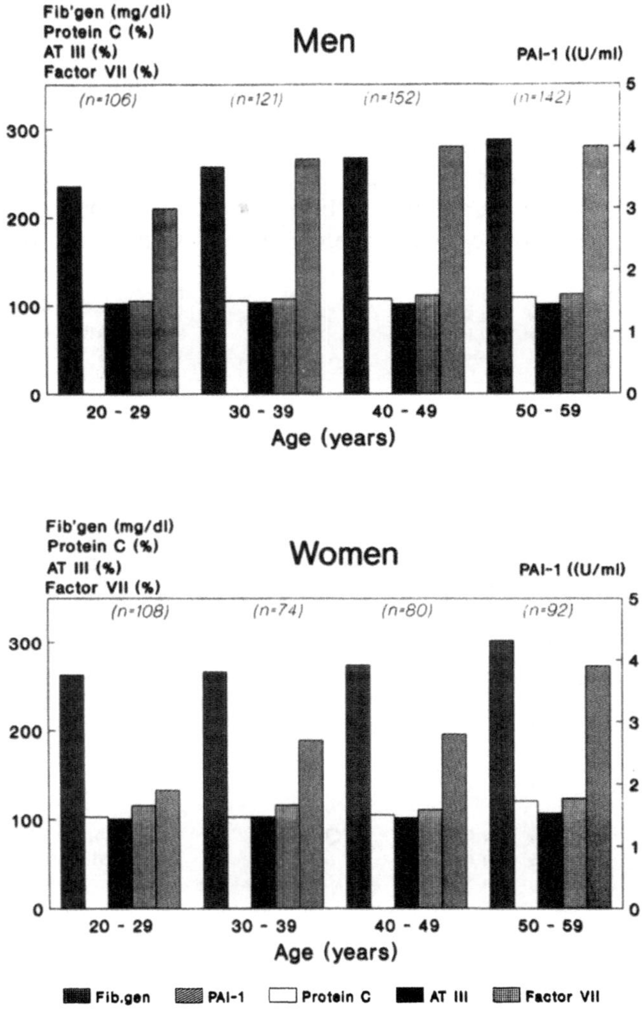

Fig. 3. Age related changes in the plasma activities of fibrinogen, protein C, antithrombin lll, factor Vlll, and PAI-1 in male and female participants of the PROCAM study. All variables significantly increase in women older than 50. Most markedly, PAI-1 increased by 50%

pective Northwick-Park-Heart study and in the PROCAM study, fibrinogen was identified as an independent CHD risk factor for men (Meade et al. 1986, Heinrich et. al 1991b). However, both studies were performed in men and therefore it remains to be established whether or not fibrinogen is also a coronary risk factor in women that becomes effective especially in elder women.

Hormones

From the results above it becomes clear that age and/or menopause importantly affect coronary risk factors in women, namely LDL-cholesterol, Lp(a), fibrinogen, and PAI-1. The most intriguing assumption to explain these changes is that estrogen exerts positive effects on these parameters. This question can, however, not directly be answered as plasma estrogen concentrations depend on the cyclus in premenopausal women and on possible replacement

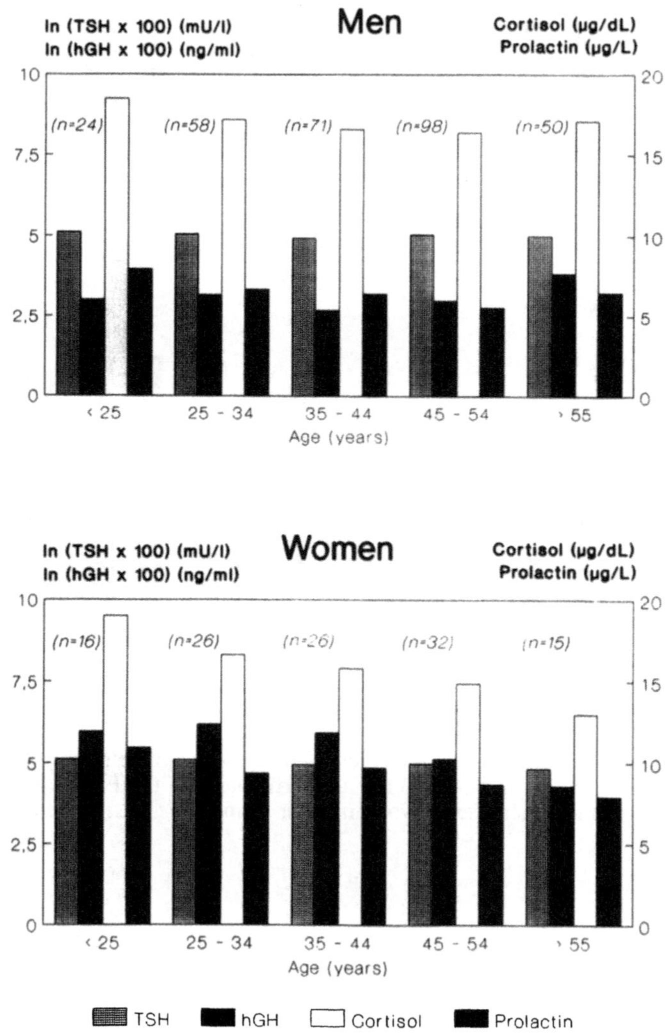

Fig. 4. Age related changes in the serum concentrations of TSH, hGH, cortisol and prolactin in male and female participants of the PROCAM study. In women concentrations of all hormones beside TSH fall throughout age, whereas only slight changes are to be observed in men

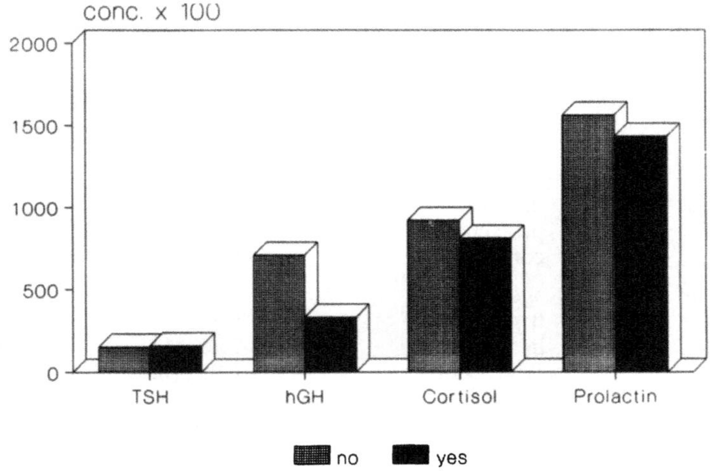

Fig. 5. Impact of menopause on hormone concentrations in women older than 40 years. Differences of hGH concentrations are significantly different ($p<0.05$, Student t-test)

therapy in postmenopausal women. Another point is that estrogens may not directly affect lipid metabolism and/or hemostasis. Therefore we studied, whether also non-sex hormones - thyreoidea stimulating hormone (TSH), human growth hormone (hGH), cortisol, and prolactin - undergo different age-related changes in men and women (Fig. 4). For data evaluation, only probands were considered whose hormone concentrations were within the normal range (TSH: 0.3–3.5 μU/ml, hGH: <40 ng/ml, cortisol <20 μg/dl, prolactin $20<x<500$ mU/l). hGH concentrations slightly increase by age in men, but severely decrease after age 45 in women. Cortisol concentrations significantly decrease in aging women but only slightly in aging men. In both sexes, TSH concentrations remain unchanged throughout life, and prolactin concentrations slightly decrease with increasing age. In women older than 40, menopause significantly discriminates between higher and lower hGH concentrations whereras the other three hormones analysed are not significantly affected (Fig 5). In both women and men hGH concentrations were negatively correlated to triglycerides and PAI-1 upon bivariate analysis (Table 3). In women but not in men, hGH concentrations positively correlated with HDL-cholesterol and apo A-I concentrations (bivariate and multivariate) (Tables 3 and 4). In women, the serum concentration of cortisol correlated negatively with LDL-cholesterol (bivariate), prolactin positively with HDL cholesterol (bivariate and multivariate). In men, prolactin affects concentrations of total cholesterol, LDL-cholesterol (also multivariate), apo B and Lp(a) negatively using bivariate analysis (Table 4). With regard to coronary risk in women, the positive correlation of both hGH and prolactin to HDL cholesterol may be important since both concentrations decrease parallel to age and since low HDL-cholesterol concentration is, besides age, the best coronary little risk

Table 3. Bivariate correlation analysis of TSH-, hGH-, cortisol-, and prolactin-serum concentrations

Parameter		Pearson correlation coefficients	
		Men (n = 303)	Women (n = 114)
ln TSH	ln Lp(a)	−0.139*	−0.023 (n.s.)
	prolactin	+0.262***	+0.056 (n.s.)
	cortisol	+0.113*	+0.020 (n.s.)
ln STH	age	+0.16*	−0.329***
	triglycerides	−0.126*	−0.244**
	ln (trigl.)	−0.184***	−0.321***
	HDL-chol.	+0.099 (n.s.)	+0.324***
	apo B	−0.081 (n.s.)	−0.228*
	apo A-I	+0.028 (n.s.)	+0.350***
	AT III	−0.137*	−0.102 (n.s.)
	PAI-1	−0.197***	−0.313***
Prolactin	age	−0.122*	−0.172 (n.s.)
	cholesterol	−0.130*	−0.105 (n.s.)
	LDL-chol.	−0.151**	−0.156 (n.s.)
	HDL-chol.	+0.027 (n.s.)	+0.225**
	ln Lp(a)	−0.119*	−0.080 (n.s.)
	apo B	+0.129*	−0.115 (n.s.)*
	factor VII	−0.154*	−0.067(n.s.)
	ln TSH	+0.262***	+0.040 (n.s.)
	cortisol	+0.228***	+0.242**
Cortisol	age	−0.079 (n.s.)	−0.353***
	LDL-chol.	−0.010 (n.s.)	−0.217*
	factor VII	−0.181**	+0.107 (n.s.)
	ln TSH	+0.113*	+0.020 (n.s.)
	prolactin	+0.228***	+0.242**

age, cholesterol, ln triglycerides, LDL-cholesterol, HDL-cholesterol, ln Lp(a), apo A-I, apo B, fibrinogen, factor VII, protein C, PAI-1, AT-III, TSH, hGH, cortisol and prolactin.
n.s. = not significant; * $p<0.05$; ** $p<0.01$; *** $p<0.001$

predictor in women (Bush et al. 1988, Corrao et al. 1990). However, these findings also adress the question why HDL cholesterol concentrations do not change in aging women.

Conclusion

In summary, the increased coronary risk in postmenopausal women has mostly been attributed to the adverse changes in the lipid profile – increased cholesterol and LDL cholesterol concentrations – due to the decrease of estrogens after menopause. Our findings point out, that also plasma concentrations of Lp(a), fibrinogen, and PAI-1 are adversely changing in aging women,

Table 4. Multiple regression analysis of TSH-, hGH-, prolactin-, and cortisol-serum concentrations

Parameter		Pearson correlation coefficients	
		Men (n = 303)	Women (n = 114)
ln TSH	ln Lp(a)	-0.139*	(n.s.)
ln STH	triglycerides	-0.184*	(n.s.)
	HDL-chol.	(n.s.)	+0.324*
	apo A-I	(n.s.)	+0.350**
	AT III	-0.137*	(n.s.)
	PAI-1	-0.197*	(n.s.)
Prolactin	LDL-chol.	-0.157*	(n.s.)
	HDL-chol.	(n.s.)	+0.225*
Cortisol	ln triglycerides	(n.s.)	+0.05**
	factor VII	-0.180**	(n.s.)
	fibrinogen	+0.087*	(n.s.)
	protein C	(n.s.)	+0.101*

multiple regression of age, menopause, OC intake, ln TSH, ln STH, prolactin, cortisol, cholesterol, ln triglycerides, LDL-cholesterol, HDL-cholesterol, ln Lp(a), apo A-I, apo B. fibrinogen, factor VII, protein C, PAI-1, and AT-III.
n.s. = not significant; * $p<0.05$; ** $p<0.01$

and that statistically significant relationships exist among these three parameters. Another important point may be derived from the sex specifically age-dependent changes of hGH, cortisol, and prolactin and their effects on coronary risk parameters of lipid metabolism in women: It is not necessarily estrogen that directly affects lipid metabolism. Other hormones may be involved, either paralleling or mediating estrogen effects.

References

Assmann G (ed) Lipid Metabolism Disorders and Coronary Heart Disease. MMV Medizin Verlag. Munich 1989.

Bush TL, Fried LP, Barrett-Connor E (1988) Cholesterol, lipoproteins, and coronary heart disease in Women. Clin Chem 34 [suppl]:B60–B70.

Corrao JM, Becker RC, Ockene IS, Hamilton GA (1990) Coronary heart disease risk factors in women. Cardiology 77 [suppl]:8–24.

Heinrich J, Sandkamp M, Kokott R, Schulte H, Assmann G (1991) Relationships of lipoprotein(a) to variables of coagulation in a healthy population. Clin Chem 37: 1950–1954.

Heinrich J, Schulte H, Balleisen L, van de Loo, Assmann G (1991) Predictive value of haemostatic variables in the PROCAM study. Thromb Haemost 65:815 (abstr.)

Loscalzo J (1990) Lipoprotein(a). A unique risk factor for atherosclerotic disease. Arteriosclerosis 10:672–679.

Meade TW, Brozowic M, Chakrabarti RR et al. (1986) Haemostatic function and ischaemic heart disease. Principal results of the Northwik Park Heart Study. Lancet II:501–505.

Different Lipoprotein Patterns in Men and Women Examined by Coronary Angiography

H. HAHMANN, D. SCHÄTZER-KLOTZ, B. KLOTZ, D. BECKER, T. BUNTE, S. KOHRING, H.-E. KELLER, and H. SCHIEFFER

Summary

By usual clinical reasons 737 men and 280 women, mean age 59 ± 10 years, were subjected in sequence to coronary angiography. We compared men and women concerning the relationship between their angiographic findings and lipoproteins as well as other risk factors.

In men and women with proven coronary artery disease (CAD) of same degree of severity we found different pattern of lipoproteins: Fe-male patients show higher levels of cholesterol (chol), low density lipoprotein chol (LDL-C) and high density lipoprotein chol (HDL-C) and lower ratios of chol/HDL-C and LDL-C/HDL-C . In women weight and age are higher, but they smoked less than men. Lipoprotein(a) [Lp(a)] levels are not different in men and women.

Discriminators of presence of CAD in female are 1st chol, 2nd smoking, 3rd high Lp(a) levels and 4th diabetes mellitus; discriminators of early manifestation are 1st family history and 2nd smoking habits (discriminant analysis). In contrast to the findings in male, lipoprotein levels in young female with CAD are lower than in older female patients.

Differences could be caused by the considerable menopausal changes of lipoprotein pattern in women.

Introduction

Various studies have shown that in women a different lipid status exists due to the hormonal situation compared to the men [5, 8]. Especially the menopause is an important turning point causing an alteration of the lipo-protein pattern the effects of which on the coronary risk are controversly discussed [2, 1, 4].

The following study war aimed at showing the differences between men and women among 1017 coronary angiographically examined patients with regard to the lipoprotein pattern and the different significance of the risk factors for CAD.

Patients and Methods

A group of 1017 patients (737 men, 280 women, mean age 59±10 years) consecutively underwent coronary angiography by JUDKINS's method. After a 12 hours fasting period, with every angiography, determinations of serum levels of the following parameters were performed:

Total cholesterol by CHOD-PAP method, LDL-C by PVS precipitation and CHOD-PAP method, HDL-C after tungstophosphoric acid precipitation, triglycerides (TG) by a fully enzymatic test, as well as Lp(a) by nephelometry (antiserum IMMUNO).

Lipoprotein levels and angiographical findings (quantified by the number of involved vessels and GENSINI's scoring system [3]) were statistically compared by MANN-WHITNEY-WILCOXON U-Test, CHI-square test and stepwise discriminant analysis [6]. The criterion for the inclusion of one variable was the significance level $p<0.05$. The thereby developed model was proved by the correctly classified cases.

Results and Discussion

Concerning the relationship between angiographic findings and lipoproteins as well as other risk factors men and women with manifested CAD show different patterns. Significant higher cholesterol, LDL-C and HDL-C levels were found in female subjects whereas the chol/HDL and LDL/HDL ratios were lower compared to men (Table 1). Despite the higher lipid- and lipoprotein levels in women the mean age for manifestation of CAD is significantly higher than for men.

Table 1. Lipid and lipoprotein levels in men and women with CAD

	Total	Coronary artery disease	
	CAD	Men	Women
CHOL [mg/dl]	245 ±49	242 ±48****	256 ± 51
LDL-C [mg/dl]	169 ±48	167 ±47*	175 ± 50
HDL-C [mg/dl]	46 ±13	44 ±12****	52 ± 15
CHOL/HDL-C	5.6± 1.8	5.8± 1.7 ****	5.3± 1.9
LDL-C/HDL-C	3.9± 1.5	4.0± 1.5 ***	3.6± 1.6
TG [mg/dl]	175 ±92	176 ±89	173 ±102
Lp(a) [mg/dl]	23 ±21	22 ±20	26 ± 22
GENSINI SCORE	44 ±33	44 ±32	43 ± 36
AGE [yrs.]	59 ± 9	58 ± 9****	62 ± 9
BROCA [%]	110 ±13	108 ±12****	113 ± 16
n	785	592	193

* $p<0.05$; ** $p<0.01$; *** $p<0.005$; **** $p<0.0005$

Table 2. Lipid and lipoprotein levels in men and women with and without CAD

	Men		Women	
	No CAD	CAD	No CAD	CAD
CHOL [mg/dl]	229 ± 45**	242 ±48	239 ±42*	256 ± 51
LDL-C [mg/dl]	152 ± 42**	167 ±47	160 ±40*	175 ± 50
HDL-C [mg/dl]	45 ± 14	44 ±12	53 ±16	52 ± 15
LDL-C/HDL-C	3.6± 1.4*	4.0± 1.5	3.2± 1.1	3.6± 1.6
TG [mg/dl]	175 ±106	176 ±89	156 ±71	173 ±102
Lp(a) [mg/dl]	16 ± 14***	22 ±20	15 ±12****	26 ± 22
Age [yrs.]	55 ± 11****	58 ± 9	60 ±10	61 ± 9
n	145	592	87	193

* p<0.05; ** p<0.005; *** p<0.001; **** p<0.0005

In the univariate analysis the important factors for the presence of CAD in women were the elevated Lp(a), cholesterol and LDL-C values. While among the male patients the age, elevated Lp(a), cholesterol and LDL-C levels as well as a higher LDL/HDL ratio were in the foreground (Table 2).

According to our former findings, the appearence of an early CAD can be considered a sign of the importance of a risk factor in men. Caused by the observation that elder patients with coronary artery disease often show lower cholesterol levels and less smoking than patients with premature atherosclerosis, we wanted to verify this relationship in female patients with CAD.

In women lipid- and lipoprotein levels increase with advancing age of CAD manifestation, following the normal age distribution (Table 3). Younger women (<53 yrs.–18th percentile of age) show significantly lower chole-

Table 3. Lipid and lipoprotein levels in men and women with early manifested CAD

	Men with CAD		Women with CAD	
	<50 yrs.+	≥50 yrs.+	<53 yrs.+	≥53 yrs.+
CHOL [mg/dl]	251 ±50*	240 ±47	241 ±45*	256 ± 51
LDL-C [mg/dl]	173 ±49	165 ±47	164 ±42	177 ± 51
HDL-C [mg/dl]	44 ±12	45 ±12	53 ±13	51 ± 15
LDL-C/HDL-C	4.2± 1.6	3.9± 1.5	3.2± 1.0	3.7± 1.7
TG [mg/dl]	199 ±90***	172 ±88	148 ±86*	178 ±104
Lp(a) [mg/dl]	32 ±29**	20 ±17	22 ±17	26 ± 23
GENSINI SCORE	41 ±31	45 ±32	34 ±30*	45 ± 36
Age [yrs.]	43 ± 5	61 ± 7	47 ± 4	64 ± 6
n	91	501	32	161

* p<0.05; ** p<0.01; *** p<0.005; + 18th Percentile

sterol- and TG levels in comparison to the older female CAD patients. On the contrary, significantly higher TG-, Lp(a)- and cholesterol levels could be noticed in younger men (<50 yrs.–18th percentile of age) with CAD.

By the stepwise multiple discriminat function analysis the significance of the risk factors is evident in the descend sequence of Table 4 and 5. In men the age played a major role for the manifestation of CAD as well as the elevated LDL-C levels, diabetes mellitus and elevated Lp(a) levels (Table 4). For the early manifestation of CAD elevated Lp(a) and TG levels were in the first place followed by the smoking habits and the family history. Hypertension and diabetes mellitus are more frequent in the elderly.

The important discriminators for the development and the early manifestation of CAD in women were elevated cholesterol levels and a positive family history, respectively, followed by the smoking habits. Levels of Lp(a) >50 mg/dl seem to be more important for older women (Table 5).

In conclusion we can state that female patients with CAD show higher lipoprotein levels and higher age compared to the male patients. The significance of risk factors differs in men and women calculated by multiple discriminant analysis.

Among other reasons these differences could be caused by the following: Contrary to female lipoprotein levels, even Lp(a), which changes marketly in the menopause [7], high lipoprotein levels in male subjects frequently indicate genetic disorders and can be considered to exist during their entire life. Therefore lipoproteins determined on the occasion of coronary angiography allow a better reconstuction of the coronary risk profile in men than in women.

Table 4. Stepwise discriminant analysis (men)
(–): lower frequence in men <50 yrs; * logarithmic transformation to produce a normal distribution

Men without CAD vs. men with CAD (angiographically examined)		
	Signif.:	Lambda:
1. Age of Patientes	p<0.00005	0.98
2. LDL-C	p<0.00005	0.96
3. Diabetes	p<0.00005	0.95
4. Lp(a) >50 mg/dl	p<0.00005	0.94
("Correctly Classified Cases": 63%)		
Men <50 yrs. vs. men ≥50 yrs. (manifest coronary artery disease)		
	Signif.:	Lambda:
1. Lp(a) >50 mg/dl	p<0.00005	0.96
2. Blood Pressure (–)	p<0.00005	0.94
3. Log TG*	p<0.00005	0.92
4. Diabetes (–)	p<0.00005	0.91
5. Smoking	p<0.00005	0.90
6. Family History	p<0.00005	0.89
("Correctly Classified Cases": 69%)		

Table 5. Stepwise discriminant analysis (women)
(–): lower frequence in women <53 yrs;

Women without CAD vs. women with CAD (angiographically examined)		
	Signif.:	Lambda:
1. Cholesterol	$p<0.01$	0.97
2. Smoking	$p<0.01$	0.95
3. Lp(a) >50 mg/dl	$p<0.0005$	0.93
4. Diabetes	$p<0.00005$	0.92
("Correctly Classified Cases": 63%)		
Women <53 yrs. vs. women ≥53 yrs. (manifest coronary artery disease)		
	Signif.:	Lambda:
1. Family History	$p<0.00005$	0.89
2. Smoking	$p<0.00005$	0.85
3. Lp(a) >50 mg/dl (–)	$p<0.00005$	0.83
("Correctly Classified Cases": 83%)		

References

1. Campos H, McNamara JR, Wilson PWF, Ordovas JM, Schaefer EJ (1988) Differences in low density lipoprotein subfractions and apolipoproteins in premenopausal and postmenopausal women. J Clin Endocrinol Metab 67:30–35.
2. Colditz GA, Willett WC, Stampfer MJ, Rosner B, Speizer FE, Hennekens CH (1987) Menopause and the risk of coronary heart disease in women. N Engl J Med 316:1105–10.
3. Gensini GG (1983) A more meaningful scoring system for determining the severity of coronary heart disease. Am J Cardiol 51:606
4. Jensen J, Riis BJ, Strom V, Nilas L, Christiansen C (1987) Long-term effects of percutaneous estrogens and oral progesterone on serum lipoproteins in postmenopausal women. Am J Obstet Gynecol 156:66–71.
5. Longcope C, Herbert PN, McKinlay S, Goldfield SRW (1990) The relationship of total and free estrogens and sex hormone-binding globulin with lipoproteins in women. J Clin Endocrinol Metab 71:67–72.
6. Norusis MJ (1988) SPSS-X advanced statistics guide,2nd edition, New York,1988.
7. Sandkamp M, Assmann G (1990) Lipoprotein (a) in PROCAM participants and young myocardial infarction survivers. In: Scanu AM, ed. lipoprotein(a): 25 years of progress. New York: Academic Press, 205–209.
8. Stefanick ML, Williams PT, Krauss RM, Terry RB, Vranizan KM, Wood PD (1987) Relationships of plasma estradiol, testosterone, and sex hormone-binding globulin with lipoproteins, apolipoproteins, and high density lipoprotein subfractions in men. J Clin Endocrinol Metab 64:723–729.

Influence of Oral and Percutaneous Estrogen Therapy on the Lipoprotein Particle Profile of Postmenopausal Women

J. M. BARD, P. AMOUYEL, and J. C. FRUCHART

Introduction

Lipids in women change towards a more atherogenic profile, basically an increase in LDL, at menopause [9]. Estrogen therapy which is used to treat the symptoms of menopause, also leads to changes in the lipid profile, most often a decrease in LDL and an increase in HDL and triglycerides

However, the use of lipoproteins defined by their density, i.e. LDL and HDL, does not take into account the protein component of the lipoproteins, which plays an important metabolic role. In this aim, lipoproteins may be distinguished according to their apolipoprotein composition. Under this concept, lipoproteins are separated into simple particles, containing only one apolipoprotein or complex particles, containing several different apolipoproteins [2]. As an example, lipoproteins containing only apo AI are called Lp AI, those containing only apo B are called Lp B. Lipoproteins containing both apo AI and apo AII are named Lp AI:AII, those containing apo B and apo E are named Lp B:E, etc...

We have shown that particles containing apo B do not behave in the same way with regard to LDL receptor binding, depending on the presence or the absence of apo E and/or apo C-III [1]. Also Lp AI was shown to be able to promote cholesterol efflux, while Lp AI:AII is not [3]. We have good evidence that Lp B: E and, to a lesser extent, Lp B:C-III may represent better risk markers for atherosclerosis than apo B or LDL cholesterol and that Lp AI may represent a better marker of protection than apo AI or HDL cholesterol [5, 12]. Therefore, we investigated the effect of 60 days of treatment of normolipemic postmenopausal women with estradiol administered either per os or percutaneously on these lipoprotein particles.

Material and Methods

Study Design

This was an open study in which 73 postmenopausal women (more than 6 months and less than 3 years), aged 45 years and older, were randomized into two groups receiving either ESTRADERM TTS 50 (estradiol 0.05 mg/d) or estradiol per os (2 mg/d), free of progestagen, for 60 days. Patients were normolipemic (cholesterol <2.60 g/l, apo B <1.30 g/l, triglycerides <1.50 g/l) and normoglycemic at inclusion time.

EDTA plasma samples were obtained at inclusion time (D0) and after 60 days of treatment (D60).

Analytical Methods

Cholesterol and triglycerides were measured in plasma by enzymatic methods (Boehringer Mannheim, FRG) adapted to an automatic centrifuge analyser. Cholesterol was also measured in the supernatant after sodium phosphotungstate precipitation (Boehringer Mannheim, FRG) to get HDL cholesterol. LDL cholesterol was estimated by the Friedewald formula (LDL cholesterol = total cholesterol – HDL cholesterol – triglycerides/5).

Apolipoproteins (apo) AI and B were analysed by immunonephelometric assays (Behring, Marburg, FRG).

Lp AI, lipoproteins containing apo AI but free of apo AII, were estimated by a differential electroimmunoassay [11] (Sebia, Issy les Moulineaux, France). Lp B:E and Lp B:C-III, particles containing both apo B and apo E or apo B and apo C-III, were analysed using a two site ELISA [8].

Statistical Analyses

An analysis of variance (ANOVA) at D0 was used to test the comparability of patients at inclusion time. The effect of each treatment was evaluated by a paired t-test. An ANOVA on the differences between D60 and D0 in each group was used to test the difference in efficacy between both therapies.

Results

Three patients on ESTRADERM TTS and two patients on estradiol per os had no blood drawn at the final visit. One patient refused to participate in the assay although she signed the informed consent form. One patient was excluded from the analysis because her menopause occured more than 3 years before the inclusion time and another because her menopause occured too recently (3 months). Two patients were excluded because they fit with one of the exclusion criteria (estradiolemia above 50 pg/ml). One patient on ESTRADERM TTS was excluded because of a bad compliance. Therefore, a total of 62 patients was retained for the final analysis, 30 patients were on estradiol per os and 32 on ESTRADERM TTS.

As shown in Table 1, cholesterol was significantly decreased (–4.9%) by estradiol per os (POS). This was due to a decrease in LDL cholesterol (–12.5%) while HDL did not change. Triglycerides were increased significantly with the oral treatment (+40.5%). The decrease in apo B (–6.6%) on

Table 1. Change in lipids, apolipoproteins and lipoproteins under estradiol per os or Estradem TTS

		Baseline	Treatment	p^1	p^2
Cholesterol	POS	2.25±0.27	2.14±0.32	0.014	0.08
	TTS	2.32±0.27	2.33±0.34	0.924	0.08
Triglycerides	POS	0.79±0.23	1.11±0.85	0.022	0.006
	TTS	0.88±0.27	0.82±0.24	0.164	0.006
LDL cholesterol	POS	1.36±0.26	1.19±0.27	0.0008	0.006
	TTS	1.44±0.25	1.44±0.32	0.919	0.006
HDL cholesterol	POS	0.74±0.23	0.75±0.18	0.638	0.86
	TTS	0.73±0.16	0.75±0.20	0.398	0.86
Apo AI	POS	1.76±0.23	1.90±0.27	0.005	0.02
	TTS	1.78±0.27	1.78±0.25	0.948	0.02
Apo B	POS	1.06±0.21	0.99±0.26	0.074	0.49
	TTS	1.12±0.18	1.08±0.23	0.133	0.49
Lp AI	POS	0.69±0.17	0.78±0.18	0.005	0.18
	TTS	0.70±0.18	0.72±0.26	0.734	0.18
LpB:E	POS	0.19±0.09	0.22±0.17	0.335	0.45
	TTS	0.22±0.10	0.22±0.14	0.915	0.45
Lp B: C-III	POS	0.10±0.05	0.11±0.07	0.545	0.06
	TTS	0.12±0.09	0.08±0.07	0.051	0.06

Means (g/l) + s.d.
1. Intra group comparison, baseline vs treated
2. Inter group comparison, POS vs TTS

estradiol per os was not significant. Apo B containing particles (Lp E: B and Lp C-III: B) were not modified either by the oral treatment. In contrast this administration route led to a significant increase in apo AI (+8%) and Lp AI (+13%).

In contrast to the oral treatment, percutaneous estradiol therapy (TTS) did not lead to any changes except for a significant decrease in Lp B:C-III.

Discussion

The analysis of variance at D0 did not reveal any differences between the two treatment groups with regard to lipid profile. This study entailed comparing the effect of two administration routes of 17 β-estradiol on the lipid and lipoprotein concentrations in postmenopausal normolipemic women. Percutaneous therapy avoids the peak concentrations of estradiol observed with the oral route and leads to a reduction in the transformation of estradiol into estrone. In this study, we observed (data not shown) an increase in the estradiol/estrone ratio only in the percutaneous therapy group. This may explain the fact that percutaneous therapy has less of an effect on the lipid profile than oral treatment.

Since a decrease in LDL cholesterol in hypercholesterolemic patients has been shown to be related to a decrease in cardiovascular morbidity [10, 7], the decreasing effect of estradiol per os on this parameter might be considered as beneficial. However, this effect is most probably counteracted by the increase in triglycerides observed in this group. Although the association between triglycerides and cardiovascular morbidity is still under discussion, the common belief is that it is better to avoid any increase in this parameter. Moreover, our recent data suggest strongly that Lp B:E and, to a lesser extent, Lp B:C-III may represent better risk markers for atherosclerosis than LDL cholesterol [5]. It is therefore intriguing to observe no effect of the oral route on these parameters. Since the plasma level of these particles is highly correlated with triglycerides [4], the increasing effect of the oral treatment on triglycerides probably explains its lack of effect on these apo B containing particles, despite the decrease in LDL cholesterol and the probable activation of the receptor specific pathway. In contrast with the oral treatment, percutaneous estradiol therapy did not lead to any changes in lipids except for a decrease in Lp B: C-III. Although the mechanism of this effect remains unclear, it is interesting to note that percutaneous therapy might have a beneficial effect on a parameter which was previously suggested to be related to cardiovascular risk [5].

Oral treatment did not lead to any increase in HDL cholesterol as would be expected [6]. However, the rather short treatment time may explain the absence of effect on this parameter.

Moreover, the increase in apo AI and Lp AI underlines the fact that the treatment may lead to structural modification of HDL particles. Since it has been suggested that Lp AI may represent the true "antiatherogenic" apo AI containing particle [5, 12], one may speculate that the action of estradiol per os on Lp AI may be of substantial benefit. Metabolic studies need to be done to determine if this increasing effect is related to overproduction or underutilization. The clinical implication of this effect would depend on this mechanism of action.

Conclusions

Percutaneous therapy with estradiol seems to have no effect on the lipid profile of normolipemic postmenopausal women. Therefore this administration route may be used to treat the menopausal symptoms of normolipemic women without any deleterious effects with regard to cardiovascular risk. Oral treatment with estradiol could have some beneficial effects on lipoproteins which might be counteracted by the increasing effect on triglycerides. Further investigation is needed to determine the interest of such a therapy in terms of cardiovascular risk.

References

1. Agnani G, Bard JM, Candelier L et al. (1991) Interaction of Lp B, Lp B:E, Lp B: C-III and Lp B:C-III: E lipoproteins with the LDL receptor on HeLa cells. Arteriosclerosis 11:1021–1029.
2. Alaupovic P (1982) The role of apoliproteins in lipid transport processes. Ric Clin Lab 12:3–21.
3. Barbaras R, Puchois P, Fruchart JC et al. (1987) Cholesterol efflux from cultured adipose cells is mediated by Lp AI particles but not by Lp AI:AII particles. Biochem Biophys Res Commun 142:63–69.
4. Bard JM, Parra HJ, Douste-Blazy P, Fruchart JC (1990) Effect of pravastatin, an HMG CoA Reductase inhibitor, and cholestyramine, a bile acid sequestrant, on lipoprotein particles defined by their apolipoprotein composition. Metabolism 39: 269–273.
5. Cambien F, Parra HJ, Arveiler D et al. (1990) Lipoprotein particles in patients with myocardial infarction and controls. Arteriosclerosis 82:III–349 (abstr. 1380).
6. De Lignieres B, Basdevant A, Thomas G et al. (1985) Biological effects of estradiol 17 beta in postmenopausal women: oral versus percutaneous administration. J Clin End Metab 62:536–541.
7. Frick MH, Ello O, Haapa K et al. (1987) Helsinki heart study: Primary-prevention trial with gemfibrozil in middle-aged men with dyslipidemia. Safety of treatment, changes in risk factors and incidence of coronary heart disease. N Engl J Med 317 :1237–245.
8. Kandoussi A, Cachera C, Parsy D, Bard JM, Fruchart JC (1991) Quantitative determination of different apolipoprotein B containing lipoproteins by an enzyme linked immunosorbent assay: apo B with apo C-III and apo B with apo E. J Immunoassay 12:305–323.
9. Kannel NB, Hjortland MC, McNamara PM, Gordon T (1976) Menopause and risk of cardiovascular disease: the Framingham Study. Ann Intern Med 85: 447–452.
10. Lipid Research Clinics Program (1984) The lipid Research Clinics Coronary Prevention Trial results: II. The relationship of reduction in incidence of coronary heart disease to cholesterol lowering. JAMA 251:365–374.
11. Parra HJ, Mezdour H, Ghalim N, Bard JM, Fruchart JC (1990) Differential electroimmunoassay on ready to use plates for human Lp AI lipoprotein particles. Clin Chem 8:1431–1435.
12. Puchois P, Kandoussi A, Fievet P et al. (1987) Apolipoprotein AI containing lipoproteins in coronaary artery disease. Atherosclerosis 68:35–40.

Effects of Different Oral Contraceptives on Apolipoprotein E

E. Malle, K. Bauer, H. Bauer, K. Ehlenz, H. Kaffarnik, and A. Steinmetz

Abstract

The present study was aimed at investigating the influence of differently composed oral contraceptives on plasma lipid and apolipoprotein levels with respect to apolipoprotein E polymorphism. Women received either monophasic preparations containing ethinylestradiol/desogestrel (n = 23), ethinylestradiol/gestodene (n = 21) or a triphasic preparation (n = 15) containing ethinylestradiol/levonorgestrel. Analysis was performed before and after 3 months on oral contraception. Whereas plasma cholesterol levels and cholesterol levels in lipoprotein subfractions, apolipoprotein A-IV and B did not change, significant increases in plasma triglycerides, apolipoprotein A-I and A-II were observed partially depending on the progestin additive. In contrast, all three different contraceptives led to a significant decrease in plasma apolipoprotein E concentrations. The mean decrease of apolipoprotein E with respect to the apo E phenotpye was 15% for apo E 3/3 and 21% for apo E 3/2 phenotypes, respectively.

Introduction

Subtle differences in carbohydrate and lipid metabolism in men and women of similar age have been attributed to the effect of endogenous sex hormones [for review see ref. 5]. In addition, estrogen/progestin oral contraceptives are responsible for further changes. Ethinylestradiol increases triglyceride levels by enhancing hepatic secretion of very low density lipoprotein apolipoprotein B (VLDL-apo B) and raises plasma high density lipoprotein(HDL)-levels mainly due to a decrease in hepatic lipase-activity. In parallel low density lipoprotein (LDL) receptor activity is stimulated by estrogen [4] whereas gestagene seems to act the opposite way [5]. Numerous studies have documented oral contraceptive-induced changes of apolipoproteins such as apo B, apo A-I, and apo A-II, whereas only scarce data refer to changes of apo E in parallel [8, 10, 12]. Apo E mediates the catabolism of apo E-rich chylomicrons and VLDL-remnants and may direct cholesteryl ester-enriched large HDL-par-

Acknowledgements: This work was supported by a grant from the Deutsche Forschungsgemeinschaft to A.S. and the Austrian FWF (P 8344 Med) to E.M.

ticles to the liver. Furthermore, apo E is involved in the conversion of VLDL remnants to LDL, in which mainly functional apo E isoforms, apo E 3 and apo E 4 are active. Three common allels ($\epsilon 2$, $\epsilon 3$ and $\epsilon 4$) code for three isoforms apo E2, E3 and E4, thus leading to six apo E phenotypes, three homozygotes (E 2/2, E 3/3, E 4/4) and three heterozygotes (E 3/2, E 4/2, and E 4/3) [13].

Apo E thus plays an important role in the metabolism of lipoproteins. The present study was aimed at investigating whether low dose oral contraceptives with different estradiol/progestin combinations would influence plasma apo E concentrations. Emphasis was put on genetic apo E phenotypes.

Materials and Methods

Subjects

Women requesting oral contraceptives at the participating outpatient site were reviewed for potential eligibility. The basic criteria for inclusion in the study were: age, 16–35 years, menstrual cycle regularity of 25–35 days; absence of contraindications to oral contraceptives, no oral contraceptive or sex steroid use within six months prior to the study. Women who consented (n = 59) received, in addition to ethinylestradiol (EE) either monophasic formulations (Table 1) containing in addition desogestrel (M-DG) or gestodene (M-GD) or a triphasic preparation containing in addition levonorgestrel (T-LN). Women in the three groups did not differ in body mass index nor in age. Baseline measurements were performed in the luteal phase between days 18 and 22 of the cycle. Blood samples were likewise taken between days 18 and 22 during the treatment period after three cycles.

Table 1. Dosage of the different oral contraceptive preparations (n = number of subjects investigated)

	composition	day
M-DG (n = 23)	30 µg ethinylestradiol 150 µg desogestrel	1–21
M-GD (n = 21)	30 µg ethinylestradiol 75 µg gestodene	1–21
T-LN (n = 15)	30 µg ethinylestradiol 50 µg levonorgestrel	1– 6
	40 µg ethinylestradiol 75 µg levonorgestrel	7–11
	40 µg ethinylestradiol 125 µg levonorgestrel	12–21

Estimation of Lipoprotein and Apolipoproteins

Blood samples were drawn after an overnight fast between 8 and 10 a.m. Plasma total cholesterol (TC) and triglycerides (TGs) were measured by commercially available test kits (Boehringer-Mannheim, Mannheim, Germany). VLDL-, LDL-, and HDL-cholesterol were estimated densitometrically after electrophoretic separation of lipoproteins (Immuno, Heidelberg, Germany). Apolipoproteins A-I, A-II, and B were quantitated by radial immunodiffusion kits (Immuno, Heidelberg, Germany). Apo E and A-IV were quantitated by electroimmunoassay (EIA) as described [12]. Phenotyping of apo E was performed from delipidated VLDL samples, subjected to isoelectric focusing on Immobiline plates (Pharmacia, Freiburg, Germany) as described [13].

Results

Table 2 displays plasma lipid and lipoprotein parameters at baseline and after three cycles of oral contraceptive administration. In all subjects investigated plasma TGs levels increased. The rise was significant in the M-GD and T-LN group. Inversely HDL-C increased significantly only in the M-DG group. Total cholesterol (TC) as well as VLDL-C (4.53±4.83 mg/dl vs 4.73±3.36 mg/dl, n=59) and LDL-C (119.6±32.6 mg/dl vs 118.9±30.9 mg/dl, n=59) were not significantly altered in any of the three groups neither when taken together from all three groups. The main apoprotein moiety of HDL, apo A-I (with the exception of the T-LN group) and apo A-II significantly increased independent of the different estrogen/progestin composition (Fig. 1), whereas apo A-IV (15.2±4.7 mg/dl vs 14.2±4.2 mg/dl, n=59) and apo B plasma concentrations remained unchanged (Fig. 1). The increase in apo A II was the most pronounced increase noted (Fig. 1).

Table 2. Plasma lipid, lipoprotein and apoprotein parameters at baseline (A) and after three cycles (B) of oral contraceptive administration. Values are given in mg/dl (mean ± SD)

parameter	drug	(A)	(B)	
TG	M-DG	86.4±43.2	104.4±44.1	NS
	M-GD	77.6±44.1	98.5±51.1	(*)
	T-LN	73.6±25.6	90.1±25.1	(#)
TC	M-DG	184.3±33.7	186.4±32.4	NS
	M-GD	180.7±32.2	175.4±36.9	NS
	T-LN	176.1±26.6	185.3±26.9	NS
HDL-C	M-DG	53.7±12.5	59.6±13.6	(#)
	M-GD	55.8±12.1	56.2±12.7	NS
	T-LN	59.1±17.7	61.1±14.7	NS

(*) = p<0.05; (#) = p<0.01; NS = non significant

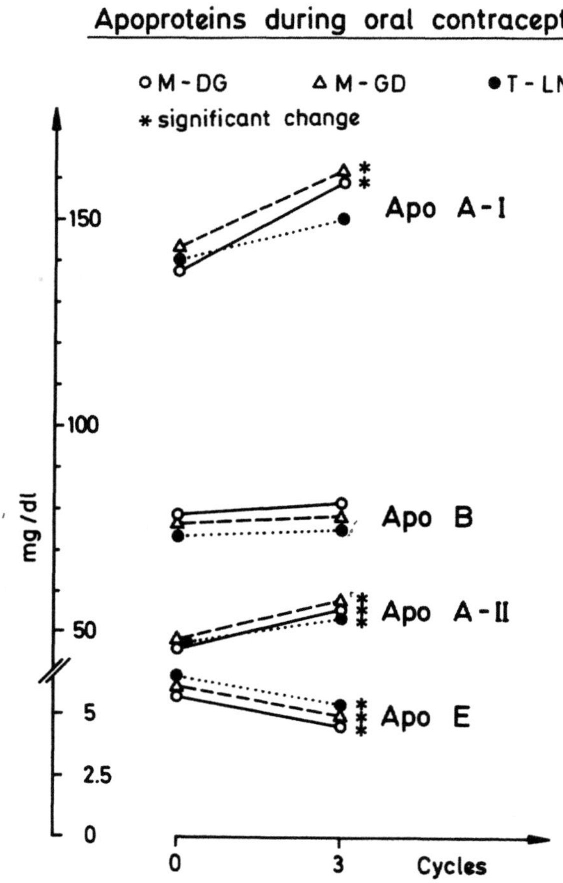

Fig. 1. Plasma apolipoprotein changes upon use of different low dose oral contraceptives in healthy young women. Apolipoproteins were measured before and after three months on medication. Mean values (given in mg/dl) are displayed

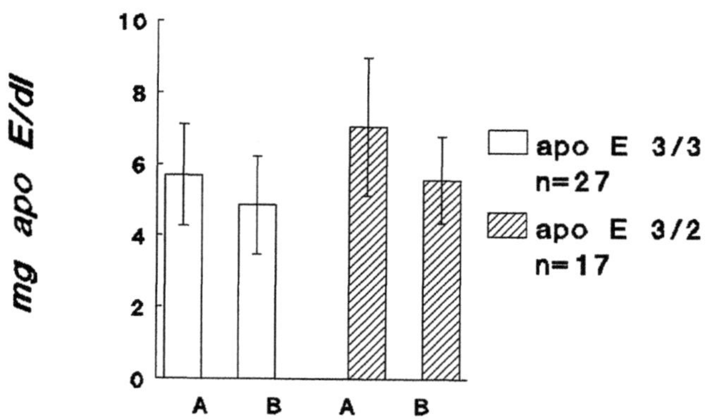

Fig. 2. Decrease of plasma apo E levels with respect to apo E phenotype before (A) and after three months (B) on oral contraceptives. Only the apo E 3/2 and apo E 3/3 phenotype groups were large enough to do statistical analysis. Women with the same apo E phenotype from the three treatment groups are taken together. Apo E concentrations were measured by EIA as described [12] and apo E phenotyping was performed from delipidated VLDL samples [13]

Figure 1 further displays the significant ($p<0.01$) decrease in plasma apo E levels after three months of oral contraceptive intake independent of the different estrogen/progestin composition. A reduction of plasma apo E levels of 15% was seen for apo E 3/3 and of 21% for apo E 3/2 (Fig. 2), respectively. The difference in the decrease in apo E between the two phenotypic groups was not significant. The apo E 4/3 group was to small ($n=3$) to do analysis upon.

Discussion

The use of oral contraceptives significantly increases the risk for thromboembolism but not for myocardial infarction and stroke. This risk mainly appears to correlate with the dose of the estrogen component but to a lesser degree with the progestin content. In addition, lipoprotein changes in women taking oral contraceptives are also critically dependent upon the dose and type of estrogen and progestin used in formulations. Estrogens in contrast to progestins increase both hepatic synthesis and secretion of HDL and VLDL. Estrogens decrease hepatic triglyceride lipase(HTGL)-activity and subsequently raise HDL_2 levels, whereas androgenic progestins increase HTGL-activity and lower HDL_2. Therefore, we studied the influence of three different low dose oral contraceptives on plasma apo E levels. Studies designed similar to our protocol with respect to the duration and the estrogen/progestin composition reported uniquely increased levels of TGs by the use of either monophasic or triphasic formulations [1, 3, 6, 7, 9, 11]. LDL-C was either found to increase [1, 3, 11] or was not significantly altered [6, 7, 9], whereas HDL-C levels were found to increase [7, 9] or even decrease [1, 3, 11]. Our findings are in parallel with studies with respect to changes of plasma levels of apo A-I and apo A-II [6, 7, 9]. Some studies reported increased levels of apo B after a three month intake of oral contraceptives [3, 6] whereas others [1, 7, 9] including our study failed to see significant changes.

Phillips and coworkers [10] reported sex-related differences in serum apo E levels (only 10–20% in the VLDL-fraction) and a significant reduction in apo E levels in females apparently due to intake of oral contraceptives. These data are in line with our findings that contraceptive steroids independently of the different estrogen/progestin composition are responsible for the decrease in plasma apo E levels. This indicates that estrogen possibly exerts beneficial effect on lipid metabolism by decreasing atherogenic β-VLDL particles and increasing hepatic uptake of chylomicron-remnants [2]. Namely, chylomicron-remnants carrying apo E 3 and apo E 4 isoforms, that permit remnant recognition are rapidly cleared from the circulation by increased hepatic uptake. These findings were deduced from observations that homozygosity for apo E 2 may be associated with normolipemic dysbetalipoproteinemia and type III hyperlipidemia, a disorder, characterized by accumulation of chylomicron- and VLDL-remnants [14]. We further could show that oral contraceptive users carrying the ϵ-3 allele had a reduction of plasma apo E levels of about 15% for apo E phenotype 3/3 and 21% for the apo E phenotype 3/2, respectively.

In summary, the present study has shown that low dose monophasic or triphasic oral contraceptives with a different estrogen/progestin composition decrease plasma apo E levels. This may in part be due to rapid removal of atherogenic remnants by hepatic receptors and perhaps specifically relate to functional apo E isoforms.

References

1. Bergink EW, Klosterboer HJ, Lund L, Nummi S (1984) Effects of levonorgestrel and desogestrel in low-dose oral contraceptive combinations on serum lipids, apolipoproteins A-I and B and glycosylated proteins. Contraception 30:61–72.
2. Berr F, Eckel RH, Kern F Jr (1986) Contraceptive steroids increase hepatic uptake of chylomicron remnants in healthy young women. J Lipid Res 27:645–651.
3. Burkman RT, Courtland Robinson J, Kruszon-Moran D, Kimball AW, Kwiterovich P, Burford RG (1988) Lipid and lipoprotein changes associated with oral contraceptive use: A randomized clinical trial. Obstet Gynecol 71:33–38.
4. Cooper AD, Nutik R, Chen J (1987) Characterization of the estrogen induced lipoprotein receptor of rat liver. J Lipid Res 28:59–68.
5. Crook D, Seed M (1990) Endocrine control of plasma lipoprotein metabolism: effects of gonadal steroids. Bailliere's Clin Endocrinol Metab 4:851–875.
6. Gevers-Leuven JA, Havekes L, van der Kooij-Pontier HA, Starmans RJH, Jansen H, Bouwhuis-Hoogerwerf ML, de Pagter HAT, Hessel LW (1984) Effect of low-dose oral contraceptives on lipoproteins and lipolytic enzymes: Differences between two commonly used preparations. Metabolism 33:1039–1042.
7. Harvengt C, Desager JP, Gaspard U, Lepot M (1988) Changes in lipoprotein composition in women receiving two low-dose oral contraceptives containing ethinylestradiol and gonane progestins. Contraception 37:565–575.
8. Kushwaha RS, Foster DM, Barrett PHR, Carey KD, Bernard MG (1991) Metabolic regulation of plasma apolipoprotein E by estrogen and progesterone in the baboon (Papio sp). Metabolism 40:93–100.
9. März W, Gross W, Gahn G, Romberg G, Taubert HD, Kuhl H (1985) A randomized crossover comparison of two low-dose contraceptives: Effects on serum lipids and lipoproteins. Am J Obstet Gynecol 153:287–293.
10. Phillips NR, Havel RJ, Kane JP (1983) Sex-related differences in the concentrations of apolipoprotein E in human blood plasma and plasma lipoproteins. J Lipid Res 24:1525–1531.
11. Powell MG, Hedlin AM, Cerskus I, Kakis G, Prudham D, Rosenrot P (1984) Effects of oral contraceptives on lipoprotein lipids: A prospective study. Obstet Gynecol 63:764–770.
12. Steinmetz A, Bauer K, Jürgensen R, Kaffarnik H (1990) Low-dose oral contraceptives lower plasma levels of apolipoprotein E. Eur J Obstet Gynecol 37: 155–162.
13. Steinmetz A (1991) Clinical implications of the apolipoprotein E polymorphism and genetic variants: current methods for apo E phenotyping. Ann Biol Clin 49: 1–8.
14. Utermann G, Jaeschke M, Menzel HJ (1975) Familial hyperlipoproteinemia type III: deficiency of a specific apolipoprotein (apo E III) in the very low density lipoproteins. FEBS Lett 56:352–356.

Effects of a New Oral Contraceptive Containing Ethinyl Estradiol and Desogestrel on Plasma Lipoproteins and Apolipoproteins

W. MÄRZ, C. JUNG-HOFFMANN, R. SIEKMEIER, H. KUHL, and W. GROSS

Summary

The effect of a new two phase oral contraceptive containing ethinyl estradiol (EE) and desogestrel (DG) (40 μg EE + 24 μg DG for 7 days, and 30 μg EE + 125 μg DG) on plasma lipoproteins and apolipoproteins has been studied in 19 healthy women. After an initial control cycle, the EE/DG combination was administered for six months, and the study was completed with a final wash-out cycle. The EE/DG combination lead to increases in total triglycerides, very low density lipoproteins, and high density lipoproteins. HDL_2-cholesterol increased significantly, whereas HDL_3-cholesterol was not affected. Apolipoproteins A-I, A-II, and B increased. While both cholesterol and triglycerides in intermediate density lipoproteins increased, there was a 20% decrease in apoE. Apolipoprotein (a) decreased by 9% and 24% after one and three months of treatment, respectively. It reverted to baseline levels in the sixth treatment cycle, and even increased in a final wash out cycle.

Introduction

There are numerous reports showing that oral contraceptives affect lipoprotein metabolism [1–3]. The effects are related to the doses of both the estrogenic and the progestogenic components. In general, estrogens increase the plasma levels of triglycerides, very low density lipoproteins, and high density lipoproteins. They reduce hepatic triglyceride lipase (HTGL) activity, [4] and high doses of estrogens lower LDL concentrations by stimulating the expression of LDL receptors in the liver [5–7].

Progestogens, in particular those with androgenic activity, have opposite effects on lipoprotein metabolism. They reduce the concentration of triglyceride-rich lipoproteins. Nortestosterone derivatives such as levonorgestrel or

---

Abbreviations

EE, ethinyl estradiol; DG, desogestrel; VLDL, very low density lipoproteins; IDL, intermediate density lipoproteins; LDL, low density lipoproteins; HDL, high density lipoproteins; C, cholesterol; TG, triglycerides; PL, phospholipids; HTGL, hepatic triglyceride lipase; OC oral contraceptive; apo, apolipoprotein.

norethisterone up-regulate HTGL, [8–10] thereby reducing HDL_2. Newer progestogens with low androgenic activity (norgestimate, gestodene, desogestrel) only slightly affect lipoprotein metabolism. Therefore, combinations containing these compounds essentially have estrogen-like effects [11, 12]. However, in some preparations the influence of androgenic progestins such as levonorgestrel may be counterbalanced by an appropriate composition of the pill [13].

We here report the effects of a new two-phase low dose contraceptive preparation containing ethinyl estradiol (EE) and desogestrel (DG) on lipoproteins and apolipoproteins.

Materials and Methods

Subjects

19 healthy women (aged 18 to 35 years) participated in the study. They had not taken hormonal contraceptives for at least three months prior to the study and were not under interfering pharmacologic therapy. Ovulation was confirmed to have taken place by the determination of serum progesterone.

Study Design

The study comprised an initial control cycle (0), six cycles (1–6) of treatment with EE/DG (40 μg EE + 24 μg DG for 7 days, and 30 μg EE + 125 μg DG), and a wash out cycle.

Fasting blood samples were obtained between days 21 and 24 of the control cycle, between days 18 and 22 of the 1st, 2nd, and 6th treatment cycle, and between days 21 and 24 of the wash out cycle.

Laboratory Procedures

Cholesterol (C), triglycerides (TG), and phospholipids (PL) were assayed enzymatically. Concentrations of VLDL, IDL LDL, HDL_2, and HDL_3 were determined in terms of their cholesterol and triglyceride content after preparative ultracentrifugation (densities 1.006 kg/L, 1.019 kg/L, 1.065 kg/L, and 1.125 kg/L). Apolipoproteins A-I, A-II, and B were measured turbidimetrically (reagents from Boehringer Mannheim, FRG). Apolipoprotein E was determined by zone immunoelectrophoresis. A polyclonal antiserum specific for apoE (Daiichi Pure Chemicals, WAK Chemie Medical, Bad Homburg) was used, and experimental conditions were as described [14]. Apolipoprotein (a) was measured with a commercial radioimmunoassay (Apolipoprotein (a) RIA 100, Pharmacia, Uppsala).

Statistic

Treatment effects were tested using WILCOXON's matched-pairs signed-rank test and FRIEDMAN's two-way analysis of variance (SPSS-X on Unisys 1100).

Results

The effects of the EE/DG preparation on plasma lipoproteins and apolipoproteins are summarized in Table 1.

Whereas total cholesterol did not change, there was an increase in plasma triglycerides by 110 % after six months intake of the OC. Total phospholipids slightly, but significantly increased, most likely due to the increase in HDL. Both VLDL-C and VLDL-TG markedly increased, and the triglyceride to cholesterol ratio in VLDL raised from 2.7 to 3.5. There also was an increase in IDL as determined in terms of both cholesterol and triglycerides. The effects on triglycerides and triglyceride-rich lipoproteins were only partially reversible in the wash out cycle.

Virtually no change was seen in LDL-C, whereas LDL-TG increased by 129%. This shows that there were major changes in the composition of the

Table 1. Effects of a two-phasic combination of ethinyl estradiol and desogestrel on mean (±SD) plasma lipoprotein and apolipoprotein concentrations [mg/dl]

Cycle	0	1	3	6	washout
cholesterol	190±30	197±28	198± 27	*209±34	*212±28
triglycerides	59±22	*99±30	*112± 27	*125±48	*86±35
phospholipids	201±34	*226±39	*224± 26	*247±35	*235±22
VLDL-C	10.1± 5.4	*13.3± 5.5	*14.9± 7.1	*16.1± 9.1	*12.7± 7.5
IDL-C	4.5± 2.2	5.8± 3.0	*6.3± 2.9	*6.3± 3.7	*6.5± 4.3
LDL-C	116.0± 27.4	111.9± 21.6	113.2± 22.9	121.7± 31.6	*134.7± 26.9
HDL-C	59.2± 11.2	*65.6± 10.4	*63.8± 10.6	*65.4± 7.7	58.3± 7.8
HDL_2-C	22.0± 8.3	*26.0± 6.8	24.0± 6.8	*25.1± 7.5	21.1± 4.7
HDL_3-C	37.2± 4.6	49.5± 7.6	39.8± 7.1	40.3± 9.5	37.2± 6.8
VLDL-TG	27.2± 17.5	*42.4± 19.2	*50.5± 21.4	*57.1± 38.0	*41.0± 20.6
IDL-TG	4.9± 3.8	8.4± 7.1	6.6± 4.9	*10.5± 5.3	6.6± 4.2
LDL-TG	15.7± 6.5	*28.8± 15.9	*36.0± 15.1	*36.4± 17.1	*26.1± 23.6
HDL-TG	11.6± 2.5	*18.9± 4.8	*18.6± 5.3	*20.7± 6.6	12.0± 3.0
apoA-I	142± 20	*167± 27	*164± 21	*174± 23	147± 17
apoA-II	33.2± 7.1	*38.9± 8.0	*40.0± 8.5	*45.6± 7.2	*36.0± 7.5
apoB	60.6± 14.7	60.5± 17.9	*69.3± 15.5	*72.5± 16.6	*71.3± 15.7
apoE	6.6± 3.8	*5.0± 1.3	*5.2± 1.6	*5.0± 1.3	6.3± 1.7
Apo(a) [U/dl] §	9.4	*8.6	7.2	9.9	*10.5

* $p < 0.05$ vs cycle 0 (WILCOXON's matched-pairs rank test); § medians are reported for apo(a) concentrations

LDL fraction. For unknown reasons, a slight, but significant elevation in LDL-C persisted after the OC had been withdrawn.

As one would expect, HDL-C increased significantly. However, a much more pronounced effect was observed on HDL-TG which increased by 80%. The increase in HDL-C was accounted for by the HDL_2 fraction, whereas HDL_3 remained unchanged. Both apo A-I and apoA-II increased by 22 and 37%, respectively. HDL-C, HDL-TG, HDL_2-C, apoA-I, and apoA-II returned to baseline levels in the wash out cycle.

Apolipoprotein B rose by 20% and remained at the elevated level after discontinuation of the OC. This is consistent with the behaviour of LDL-cholesterol in the wash out cycle and may indicate that some OC induced changes in plasma lipoproteins may persist a certain period of time.

Interestingly, apoE levels decreased by 20% during treatment and returned to pre-treatment levels in the wash out cycle. Moreover, the median apo(a) concentrations decreased by 9% and by 24% in the first and third treatment cycle, respectively. In the sixth treatment cycle it slightly exceeded the baseline value, whereas in the final wash out cycle it was by 11% higher than in the control cycle.

Discussion

The EE/DG combination used in this study is a lower dose contraceptive. Nevertheless, it influenced the plasma lipoprotein pattern. The effects demonstrate a preponderance of the estrogenic over the progestogenic component of the preparation. Estrogens have repeatedly been shown to increase triglycerides, triglyceride-rich lipoproteins, HDL and its subfractions, as well as apoA-I, apoA-II, and apoB. Essentially, these changes are brought about by an enhanced secretion of triglyceride-rich lipoproteins, a reduction in HTGL, and possibly an increased synthesis of apoA-I [15].

An interesting finding in this study was that the OC caused a significant decrease in apoE. This is in good agreement with another report on apoE in females taking oral contraceptives [16]. ApoE is a major ligand for cellular lipoprotein receptors. It mediates the uptake of VLDL and chylomicron derived remnants into the liver. These remnants are enriched in cholesteryl esters and in apoE, and the apoE plasma concentration reflects the level of remnants in the circulation. At least VLDL remnants are catabolized through the interaction of apoE with hepatic LDL-receptors. Therefore, stimulation of LDL-receptor activity, for instance with HMG-CoA-reductase inhibitors [17], lowers remnant concentrations. In animals and in cell culture studies estrogens have been demonstrated to increase the expression of LDL-receptors [18–20], and one explanation for the reduction in apoE may therefore be an enhanced clearance of remnants. This would then resemble the hypolipidemic response to estrogens observed in patients with type III hyperlipoproteinemia [5]. However, since in this study, the amounts of cholesterol and triglycerides in VLDL and in particular in IDL (which contains most remnant particles) increased, it appears unlikely that the remnant concentrations had actually decreased. In addition, LDL-C and apoB remained unaffected, and this is also inconsistent with a marked up-regulation of LDL-receptors.

Another explanation for the reduction in apoE would be an inhibition of its synthesis and secretion in the liver. Triglyceride-rich particles which are relatively depleted in apoE may exhibit poor binding to LDL-receptors. Along with the reduction in the HTGL this may lead to the accumulation of IDL in our probands. In order to prove this hypothesis, apoE should be determined in lipoprotein fractions rather than in whole plasma. This will clarify whether the decrease of apoE indeed occurs in apoB containing lipoproteins or in those containing apoA-I.

Our results also show, that Lp(a) concentrations may be influenced by sex hormones. This is remarkable because Lp(a) concentrations are deemed insensitive to most exogenous factors and drugs that normally affect lipoprotein metabolism. Moreover, the data are in good agreement with a previous communication showing that conjugated equine estrogens reduce Lp(a) in postmenopausal women [21]. However, the biphasic behaviour of Lp(a) concentrations in our study (decrease within the first three months followed by an increase) is not easily explained. In addition, it is not entirely clear whether the changes are attributable to the estrogenic or to the progestogenic component of the preparation. Androgenic steroids are known to reduce Lp(a) [22, 23], but the androgenic activity of desogestrel is low, and most effects observed in this study relate to the estrogen rather than to the progestin. Therefore, the estrogen component should be responsible for the reduction in Lp(a).

It is not known whether estrogens lower Lp(a) by directly modulating synthesis or secretion of apo(a), or whether indirect mechanisms are active. One such indirect mechanism would involve HTGL. In plasma, HTGL participates in the conversion of VLDL remnants to LDL. Presently, it cannot be ruled out that Lp(a) is assembled in the circulation rather than being secreted as the mature Lp(a):B complex. If the latter held true, then HTGL might have a role in Lp(a) turnover by providing precursor LDL particles, and low HTGL activity might be rate-limiting to the production of Lp(a).

In summary, the OC used in this study had estrogen-like effects on lipoprotein metabolism. The changes in apoE and in apo(a) are interesting phenomena deserving future studies which, in turn, will provide further insights into the hormonal control of lipoprotein metabolism.

Acknowledgements

The authors thank Mrs. Sabine Cezanne and Mrs. Bettina Donnerhak for expert technical assistance.

References

1. Fotherby K (1985) Oral contraceptives, lipids, and cardiovascular disease. Contraception 31:367–394.
2. Knopp RH (1988) Cardiovascular effects of endogenous and exogenous sex hormones over a woman's lifetime. Am J Obstet Gynecol 158:1630–1643.
3. Tikkanen MJ (1990) Role of plasma lipoproteins in the pathogenesis of atherosclerotic disease, with special reference to sex hormone effects. Am J Obstet Gynecol 163:296–304.
4. Applebaum DM, Goldberg AP, Pykalisto OJ, Brunzell JD, Hazzard WR (1977) Effect of estrogen on post-heparin lipolytic activity: selective decline in hepatic triglyceride lipase. J Clin Invest 59:601–608.
5. Kushwaha RS, Hazzard WR, Gagne C, Chait A, Albers JJ (1977) Type III hyperlipoproteinemia: paradoxical hypolipidemic response to estrogen. Ann Int Med 87:517–525.
6. Stuyt PMJ, Demacker PNM, van't Laar A (1986) A study of the hypolipidemic effect of estrogen in type III hyperlipoproteinemia. Horm Metabol Res 18: 607–610.
7. Tikkanen MJ, Nikkilä EA, Vartiainen E (1978) Natural oestrogen as an effective treatment for type II hyperlipoproteinemia in postmenopausal women. Lancet II: 490–491.
8. Tikkanen MJ, Nikkilä EA, Kuusi T, Sipinen S (1981) Different effects of two progestins on plasma high density lipoprotein (HDL_2) and postheparin plasma hepatic lipase activity. Atherosclerosis 40:365–369.
9. Lipson A, Stoy DB, LaRosa JC, Muesing RA, Cleary PA, Miller VT, Gilbert PR, Stadel B (1986) Progestins and oral contraceptive-induced lipoprotein changes: A prospective study. Contraception 34:121–134.
10. Tikkanen MJ, Nikkilä EA, Kuusi T, Sipinen S (1982) High density lipoprotein-2 and hepatic lipase: Reciprocal changes produced by estrogen and norgestrel. J Clin Endocrinol Metab 54:1113–1117.
11. März W, Jung-Hoffmann C, Heidt F, Gross W, Kuhl H (1990) Changes in lipid metabolism during 12 months of treatment with two oral contraceptives containing 30 μg ethinyl estradiol and 75 μg gestodene or 150 μg desogestrel. Contraception 41:245–258.
12. Kuhl H, März W, Jung-Hoffmann C, Heidt F, Gross W (1990) Time-dependent alterations in lipid metabolism during treatment with low-dose oral contraceptives. Am J Obstet Gynecol 163:363–369.
13. März W, Groß W, Gahn G, Romberg G, Taubert HD, Kuhl H (1985) A randomized crossover comparison of two low-dose contraceptives: effects on serum lipids and lipoproteins. Am J Obstet Gynecol 153:287–293.
14. März W, Gross W (1983) Quantification of human serum lipoprotein Lp(a): zone immunoelectrophoresis assay, a new sensitive method, as compared to electroimmuno assay. Clin Chim Acta 134:265–279.
15. Schaefer EJ, Foster DM, Zech LA, et al. (1983) The effects of estrogen administration on plasma lipoprotein metabolism in premenopausal females. J Clin Endocrinol Metab 57:262–267.
16. Malle E, Bauer K, Bauer H, Ehlenz K, Kaffarnik H, Steinmetz A (1993) Effects in different oral contraceptives on apolipoprotein E. In: Steinmetz A, Schneider J, Karffarnik H (eds) Hormones in lipoprotein metabolism. Springer, Berlin Heidelberg New York, pp 137–142.
17. März W, Grützmacher P, Paul D, Siekmeier R, Schoeppe W, Groß W: Treatment of primary hypercholesterolemia with lovastatin lowers intermediate density lipoproteins, submitted for publication.
18. Arbeeny CM, Eder HA (1980) Effects of 17 alpha-ethinyl estradiol on the serum lipoproteins of cholesterol fed diabetic rats. J Biol Chem 255:10547–10550.

19. Kovanen PT, Brown MS, Goldstein JL (1979) Increased binding of low density lipoprotein to liver membranes from rats treated with 17 alpha-ethinyl estradiol. J Biol Chem 254:11367–11373.
20. Ma PTS, Yamamoto T, Goldstein JL, Brown MS (1986) Increased mRNA for low density lipoprotein receptor in livers of rabbits treated with 17 alpha ethinyl estradiol. Proc Natl Acad Sci USA 83:792–796.
21. Soma M, Fumagalli R, Paoletti R, et al. (1991) Plasma Lp(a) concentration after estrogen and progestagen in postmenopausal women. Lancet 337:612.
22. Albers JJ, Taggard HM, Applebaum-Bowden D, et al. (1984) Reduction of lecithin-cholesterol acyltransferase, apolipoprotein D, and the Lp(a) lipoprotein with the anabolic steroid stanozolol. Biochim Biophys Acta 795:293–296.
23. Baumstark MW, v. Stein Th, Jakob E, Luley C, Berg A, Keul J (1993) Testerone lowers serum concentrations of lipoprotein Lp(a). In: Steinmetz A, Schneider J, Kaffarnik H (eds) Hormones in lipoprotein metabolism. Springer, Berlin Heidelberg New York, pp 165–168.

Anabolic Steroids do not Change mRNA Levels and Protein Secretion of Apolipoprotein A-I and B-100 in HepG2 Cells

K. J. LACKNER, W. MÄRZ, G. WOLTER, H. SARTOR, and W. GROSS

Summary

The effects of two anabolic steroids, stanozolol and methenolone, on gene expression and protein secretion of apolipoprotein A-I and apolipoprotein B-100 were analyzed in the human hepatoma cell line HepG2. Concentrations of mRNA for apolipoproteins A-I and B-100 were determined by quantitative slot blot hybridization. The amount of apolipoproteins secreted into the media were measured by zone immuno electrophoresis or enzyme linked immunosorbent assay. At concentrations of 10^{-9} M to 10^{-7} M no effect on mRNA concentration or apolipoprotein secretion could be detected. At higher concentrations of the anabolic steroids non-specific toxic effects on the HepG2 cells were observed. It is concluded that the known effects of anabolic steroids on lipoprotein metabolism *in vivo*, i.e. lowering of high density lipoprotein cholesterol and apolipoprotein A-I and increase of low density lipoprotein cholesterol and apolipoprotein B-100, is not initiated by significant changes in synthesis and secretion of the respective apolipoproteins.

Introduction

Anabolic steroids induce profound changes in lipoprotein metabolism. The two major effects on plasma lipoproteins are a decrease of apolipoprotein A-I (apoA-I) and high density lipoprotein (HDL) cholesterol and an increase in apolipoprotein B-100 (apoB-100) and low density lipoprotein (LDL) cholesterol [1–3]. These changes in lipoprotein levels lead to a lipoprotein profile that has been shown in several epidemiologic studies to be associated with a considerably higher risk for cardiovascular disease [4–6]. Recently, there have been reports of sudden cardiac death in athletes using anabolic steroids [7]. Thus, there may be a link between anabolic steroids, lipoproteins, and the development of premature coronary heart disease in athletes.

Abbreviations

apo – apolipoprotein; LDL – low density lipoprotein; HDL – high density lipoprotein; HTGL – hepatic triglyceride lipase

The mechanism by which anabolic steroids affect lipoprotein metabolism is not yet entirely clear. *In vivo* studies suggest an effect on synthesis as well as on catabolism of HDL and its major apolipoproteins apoA-I and apoA-II [8]. Furthermore, there is a profound increase in the activity of hepatic triglyceride lipase (HTGL) under the influence of androgenic steroids. [8, 9]. In one *in vitro* investigation testosterone has been shown to reverse the increase in apoA-I secretion brought about by estrogens. The mechanism causing this inhibition of the estrogen effect was not further analyzed [10]. In the present study, we have investigated the effect of two commonly used anabolic steroids – stanozolol and methenolone – on the synthesis and secretion of apoA-I and apoB-100 in the human hepatoma cell line HepG2.

Materials and Methods

mRNA coding for apoA-I and ApoB was determined by quantitative slot blot hybridization. RNA was prepared by CsCl centrifugation. RNA-samples were blotted in triplicate on Gene-Screen-Membranes and hybridized with radiolabelled cDNA probes. The amount of radioactive probe bound to the RNA was quantitated in a scintillation counter after the bands had been excised. Standard curves were obtained from serial dilutions of RNA. The radioactivity bound increased linearly with the RNA over the range of interest (0.5–5 μg RNA for apoB-100).

Secretion of apolipoproteins A-I and B was measured by zone immunoelectrophoresis and by a non-competitive sandwich enzyme immunoassay.

HepG2 cells were grown in RPMI 1640 supplemented with 10% fetal calf serum. In all incubation experiments, media were changed 24 hrs before steroids were added to allow the cells to adapt to the fresh media.

Results

In preliminary experiments, the concentration of anabolic steroids to be used was defined. The physiologic concentration of testosterone is approximately 10^{-9} M. Concentrations from 10^{-11} M to 10^{-5} M were tested in the cell culture. At the highest concentration of both stanozolol and methenolone cytotoxic effects were obvious. Thus, in all further experiments either 10^{-7} M or 10^{-9} M steroids were used.

To further analyze the effect of these steroid concentrations on cellular function, total cellular protein was determined at 0, 24, 48, and 72 hrs after addition of hormone. There was no significant difference between controls and hormone treated cells.

Figure 1 shows the relative amounts of apoA-I (A) and apoB-100 (B) mRNA determined by slot blot hybridization. The data represent means from two experiments. No significant effect of either anabolic steroid could be detected. The incubation conditions were chosen such that the media were changed 24 hrs prior to addition of the hormones. This should minimize effects of the addition of fresh media. We and others have observed that there is a rapid increase in both apoA-I and apoB-100 message immediately after a change of media [16]. After 24 hrs, when the hormones were added, there is still an increase in the message compared to baseline, which is no longer detectable after 48 hrs. It is assumed that this explains the slight reduction in relative apoA-I and apoB-100 mRNA-levels within the first 24 hrs of incubation with hormones. This reduction was observed in hormone treated as well as in control cells. 24 hrs after the addition of hormone, i.e. 48 hrs after the change of media no further significant change was observed.

These data show that there is no change in mRNA levels. However, there might be a change in secretion independent from the respective message levels. To investigate this question, the amounts of apolipoproteins accumulating in the media have been determined.

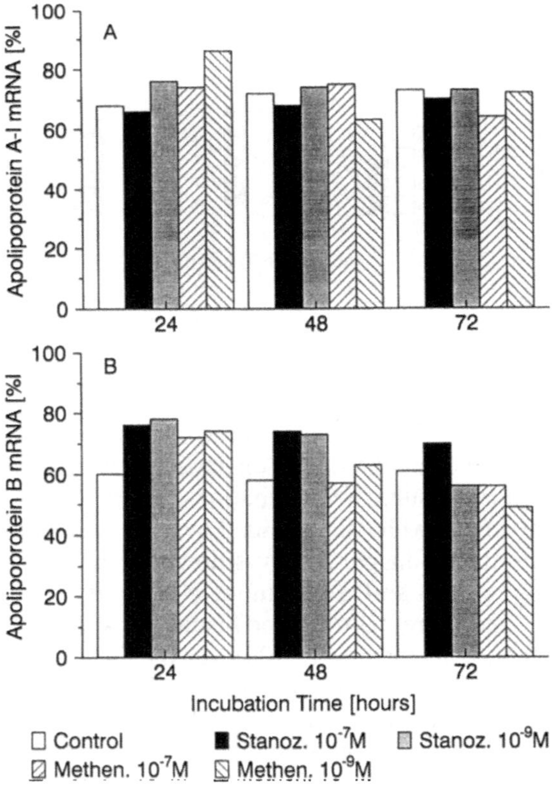

Fig. 1. Determination of apoA-I and apoB mRNA. The relative mRNA concentrations were normalized to β-actin mRNA, and the amount immediately before hormone addition (t_0) was taken as reference. The other time points (24, 48, 72 hrs) were expressed as percentages of t_0. The data are the average of two experiments. Panel A shows the results for apoA-I, and panel B for apoB. There is a slight decrease of apoA-I and apoB mRNA within the first 24 hrs after hormone addition (see text). No systematic difference between control cells and hormone treated cells could be detected

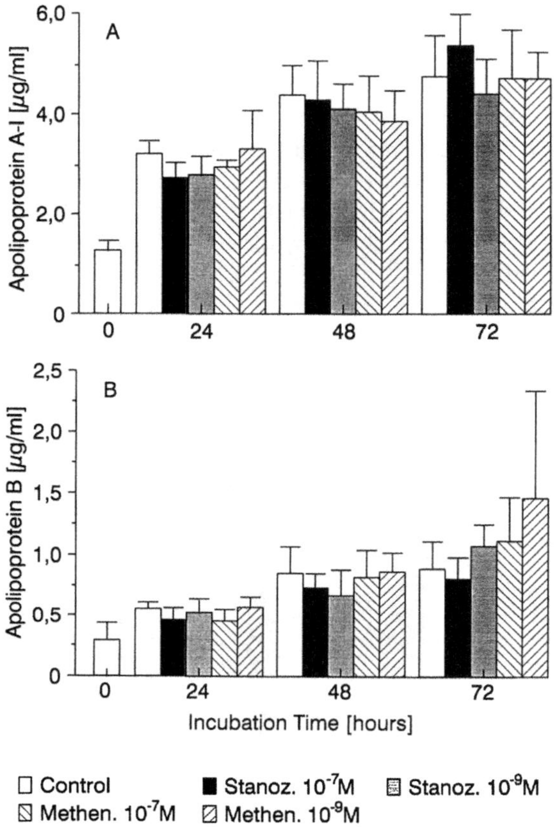

Fig. 2. Apolipoprotein secretion into the media. The data are averages from 6 experiments with standard deviations. Again panel A represents the data for apoA-I, and panel B the data for apoB. Hormone was added at t_0. Both apolipoproteins are secreted over the length of the observation period. However, there is a slight decrease in net secretion after 24 hrs. There is no difference between control and hormone treated cells. There is much larger variability in apoB secretion from one experiment to the other than in apoA-I. This is reflected in larger standard deviations. This problem becomes particularly obvious after 72 hours

Figure 2 A summarizes the effects of incubations of HepG2 cells with stanozolol or methenolone on apoA-I secretion into the media. The data represent the results of 6 incubations. There is a continuous increase in apoA-I in the media which is sustained over 72 hrs. However, the net amount secreted into the media per 24 hours slowly decreases. There is no difference between the controls and the hormone incubated cells.

In Figure 2 B the same results for apoB-100 are shown. Again there is an increase in total apoB-100 over the 72 hrs, which is faster in the first 24 hrs than in the final 24 hrs. The anabolic steroids do not have an effect on apoB-100 secretion by the HepG2 cells.

Thus, secretion of the two apolipoproteins A-I and B-100 is not affected by incubation with either stanozolol nor methenolone.

Discussion

The results presented here show that anabolic steroids do not have a major effect on secretion of apoA-I and apoB-100 in HepG2 cells. HepG2 cells have been shown to secrete the majority of relevant apolipoproteins as well as lipoprotein particles. In addition, they regulate their lipid metabolism in a way very similar to normal hepatic lipoprotein metabolism [17–19]. Thus, it is reasonable to assume that the effects seen in this cell culture system are similar to the *in vivo* effects of anabolic steroids on hepatocytes in man.

The data presented indicate that the reduction in HDL-cholesterol and the increase of LDL-cholesterol observed in man under treatment with anabolic steroids are not caused by a direct effect on synthesis or secretion of the two major apolipoproteins in these lipoprotein density fractions.

The lack of a direct effect does not rule out a secondary, additional effect on apolipoprotein secretion. That means that the decrease in synthetic rate of apoA-I which has been described in one turnover study [8] could well be brought about by changes in intravascular lipoprotein metabolism which are not reproduced in the cell culture system. However, this sequence of events is unlikely to be initiated by a reduction in apoA-I secretion and an increase in apoB-100 secretion.

The only consistent direct effect of anabolic steroids on lipoprotein metabolism is perhaps the induction of hepatic triglyceride lipase (HTGL) [8, 9]. This induction would have at least two effects. First, the rate of conversion of IDL to LDL should be increased. Second the hydrolysis of triglycerides and phospholipids on HDL should be increased. These two effects might account for the reduction in HDL_2-cholesterol and an increase in LDL-cholesterol. However, it appears unlikely that the profound reduction in HDL-cholesterol can be explained solely by the induction of HTGL. One explanation might be that the doses used by some athletes may already exert nonspecific toxic effects on the liver that lead to a reduction in HDL production. Since we tried to exclude nonspecific toxicity in our experiments, such an effect might have been missed. Furthermore there may be changes in other components of lipoprotein metabolism, e.g. in receptor activity or synthesis of lipid exchange proteins such as cholesterol ester transfer protein.

Since the androgen effect plays a role in the differences in lipoprotein levels between men and women, which account at least for some of the difference in the prevalence of cardiovascular disease in men and women, the mechanisms of action of androgenic steroids on lipoprotein metabolism are highly important. Our data exclude any major direct effect on the synthesis and secretion of apoA-I and apoB-100.

Acknowledgements

The authors thank Mrs. Sabine Cezanne and Mrs. Bettina Donnerhak for expert technical assistance. Part of this study was supported by the Deutsche Forschungsgemeinschaft, grant La 499/2-1.

References

1. Alèn M, Rahkila P, Marniemi J (1985) Serum lipids in power athletes self-ad ministering testosterone and anabolic steroids. Int J Sports Med 6:139–144 .
2. Webb OLs Laskarzewski PM, Glueck CJ (1984) Severe depression of high-den sity lipoprotein cholesterol levels in weight lifters and body builders by self-administered exogenous testosterone and anabolic-androgenic ste roids. Metabolism 33: 971–975.
3. Taggart HM, Applebaum-Bowden D, Haffner S, et al. (1982) Reduction in high density lipoproteins by anabolic steroid (stanozolol) therapy for postme nopausal osteoporosis. Metabolism 31:1147–1152.
4. Castelli WP, Garrison RJ, Wilson PWF, et al. (1986) Incidence of coronary heart disease and lipoprotein cholesterol levels. The Framingham Study. J Am Med Assoc 256:2835–2838.
5. Study Group, European Atherosclerosis Society (1987) Strategies for the pre vention of coronary heart disease: a policy statement of the EAS. Eur Heart J 8: 77–88.
6. Maciejko JJ, Holmes DR, Kottke BA, et al. (1983)Apolipoprotein A-I as a marker of angiographically assessed coronary artery disease. N Engl J Med 309: 385–389.
7. McNutt RA, Ferenchick GS, Kirlin PC, Hamlin NJ (1988) Acute myocardial in farction in a 22-year-old world class weight lifter using anabolic steroids. Am J Cardiol 62:164.
8. Haffner SM, Kushawa RS, Foster DM, Applebaum-Bowden D, Hazzard WR (1983) Studies on the metabolic mechanism of reduced high density lipo proteins during anabolic steroid therapy. Metabolism 32:413–420.
9. Tikkanen MJ, Nikkila EA (1987) Regulation of hepatic lipase and serum lipo proteins by sex steroids. Am Heart J 113:562–568.
10. Tam S-P, Archer TK, Deeley RG (1984) Effect of estrogen on apolipoprotein secretion by the human hepatocarcinoma cell line HepG2. J Biol Chem 260: 1670–1675.
11. Law SW, Brewer HB jr (1984) Nucleotide sequence and the encoded amino acids of human apolipoprotein A-I mRNA. Proc Natl Acad Sci USA 81:66–70.
12. Lackner KJ, Monge JC, Gregg RE, et al. (1986) Analysis of the apolipoprotein B gene and messenger ribonucleic acid in abetalipoproteinemia. J Clin Invest 78: 1707–1712.
13. Chirgwin JM, Przybyla AE, MacDonald RJ, Rutter WJ (1979) Isolation of bio logically active ribonucleic acid from sources enriched in ribonuclease. Biochemistry 18:5294–5299.
14. März W, Gross W (1983) Quantification of human serum lipoprotein Lp(a): zone immunoelectrophoresis assay, a new sensitive method as compared to electroimmuno assay. Clin Chim Acta 134:265–279.
15. Tijssen P (1985) Practice and theory of enzyme immunoassays. Amsterdam, New York, Oxford: Elsevier 238 ff.
16. Monge JC, Hoeg JM, Law SW, Brewer HB jr (1986) Low density lipoproteins induce coordinate regulation of human apolipoprotein A-I and B-100 biosynthesis in HepG2 cells. Clin Res 34:443a.
17. Knowles BB, Howe CC, Aden DP (1980) Human hepatocellular carcinoma cell lines secrete the major plasma proteins and hepatitis B surface antigen. Science 209: 497–499.
18. Zannis VI, Breslow JL, SanGiacomo TR, Aden DP, Knowles BB (1981) Characterization of the major apolipoproteins secreted by two human hepatoma cell lines. Biochemistry 20:7089–7096.
19. Erickson SK, Fielding PE (1986) Parameters of cholesterol metabolism in the human hepatoma cell line HepG2. J Lipid Res 27:875–883.

VLDL ApoB-100 and ApoE Metabolism in Male and Female Subjects Utilizing the Stable Isotope Technique

J. R. SCHAEFER, D. J. RADER, R. E. GREGG, T. FAIRWELL, L. A. ZECH, M. R. KINDT, and H. B. BREWER, JR.

Summary

It is well known that triglyceride levels in male and female subjects are different. Women have lower triglyceride levels than men. In order to investigate the metabolic basis for this we performed stable isotope kinetic studies in four healthy normal volunteers, two female and two male subjects. All probands received a single bolus injection of 10 mg/kg bodyweight of $^{13}C_6$ phenylalanine. Multiple blood draws were performed and free plasma amino acids as well as very low density lipoprotein fractions (VLDL) were isolated. The VLDL apolipoproteins (apo), apoB-100 and apoE, were isolated by SDS Gel. The $^{13}C_6$ phenylalanine tracer enrichment in the free amino acids and in VLDL apolipoproteins were determined by gas chromatograph – mass spectrometry (GC-MS). As the rate of tracer uptake into individual proteins is determined by the protein's *in vivo* metabolism, we were able to calculate the fractional catabolic rates (FCR) and production rates (PR) for VLDL apoB-100 and apoE. Interestingly the two female subjects had similar PR but increased FCR values for VLDL apoB-100 in comparison to the two male subjects. This data indicates that the lower triglyceride levels in females are caused by an increased FCR rather than a decreased PR. For VLDL apoE this small group did not show any relevant differences between the male and female subjects.

Introduction

Triglyceride levels in females are lower than in male subjects [10]. However the metabolic basis for this is not quite clear. *In vivo* kinetic studies can provide an answer and we therefore applied a newly developed stable isotope technique to investigate the metabolic basis for lower triglyceride and VLDL apoB-100 levels in females. An amino acid labeled with a stable isotope (^{13}C or ^{2}H) is used to endogenously label all proteins. The labeled amino acid may be administered either as a primed constant infusion, thus rapidly achieving a steady state isotopic enrichment of the precursor amino acid, or as a bolus injection, resulting in a pulse labeling of proteins. In this paper, we report the endogenous labeling of apolipoproteins using the bolus injection approach with $^{13}C_6$ phenylalanine in four normal volunteers. To investigate the trigly-

Table 1. Patient characteristics and plasma lipid values

Subject	Sex	Age	Chol	TG	LDL-Chol	HDL-Chol	ApoA-I	ApoB	ApoE
			mg/dl						
1	f	21	143	55	83	49	115	121	4
2	f	22	163	74	86	62	135	115	5
3	m	21	147	102	81	46	100	126	4
4	m	23	156	119	97	35	88	132	6

ceride kinetics we analyzed the structural protein for the VLDL metabolism, the apoB-100 and in addition apoE as an other critical apolipoprotein for triglyceride metabolism.

Study Subjects

We investigated the in vivo kinetics of VLDL apoB-100 and apoE in four normal subjects, two males and two female normal volunteers. They all were healthy, were on no lipid affecting drugs, had no lipid abnormalities and their apoE phenotype was apoE 3/3. The lipid values are illustrated in table 1. The study protocol was approved by the Internal Review Board of the National Heart Lung and Blood Institute and informed consent was given by each of the participants.

Methods

The amino acid used was a custom-synthesized six-times ^{13}C labeled L-phenylalanine, with the six labeled carbons all positioned in the aromatic ring (Cambridge Isotope Laboratories, Woburn, MA, USA). Phenylalanine was chosen in part because it undergoes an irreversible conversion to tyrosine in the liver, thus providing a second amino acid which can be used to generate a second kinetic curve for the $^{13}C_6$ tyrosine amino acid and protein kinetics (data not shown). Multiple-labeled amino acids were utilized in order to shift the labeled amino acid peak well out of the range of the peak due to the natural abundance of heavy isotopes. Multiple-labeling therefore significantly increases the accuracy of the mass spectrometric detection of the tracer, since background levels of six-times labeled phenylalanine are virtually undetectable. Labeled amino acids were dissolved in physiological salt solution, sterile filtered, and tested for pyrogenicity and sterility prior to the study. The $^{13}C_6$-phenylalanine was administered as a bolus of 10 mg/kg. Blood (20 ml EDTA) was drawn before injection, at 10 minutes and multiple time points after injection. Blood samples were drawn into tubes containing EDTA at a final concentration of 0.1%, plasma was separated by low-speed centrifugation, and sodium azide and aprotinin were added at final concentrations of 0.05% and 200 KIU/ml, respectively.

Plasma amino acids and VLDL apoB-100 and apoE were isolated, derivatized and analyzed on GC-MS corresponding to an earlier published protocol [9]. Briefly, free plasma amino acids were isolated from 0.5 ml plasma by cation exchange column chromatography [14]. The samples were derivatized to the N-heptafluorobutyryl isobutyl ester [5]. Lipoprotein subfractions were isolated by sequential ultracentrifugation of 5 ml plasma [2]. Density subfractions were immediately dialyzed against 0.01 M NH_4HCO_3, shell-frozen on dry ice, and lyophilized.

ApoB-100 and apoE was isolated from VLDL by preparative discontinuous gradient SDS-PAGE (5/15% acrylamide) using a modification of a previously published method [6]. After electrophoresis, gels were stained, destained and the apoB-100 and apoE bands were cut from the gels. The protein-containing gel slices were hydrolyzed in 6N HCl for 24 h at 110 °C under nitrogen vacuum. After removal of HCl the samples were applied to cation exchange columns. The protein-derived amino acids were derivatized to the N-heptafluorobutyryl isobutyl esters for analysis on GC-MS.

The isotopic enrichment (IE), defined as the ratio of labeled to unlabeled and labeled amino acid, was quantitated by gas chromatograph-mass spectrometry on a Finnigan MAT 4500 (Finnigan MAT, San Jose, CA, USA) in the chemical ionization mode, using isobutane as the reagent gas. The selected positively charged ions of 418 m/z for phenylalanine and 424 m/z for $^{13}C_6$-phenylalanine, 630 m/z for tyrosine and 636 m/z for $^{13}C_6$-tyrosine were monitored. Three measurements were made of each protein at each time point.

Analytical Methods

VLDL ApoB was quantitated by ELISA. VLDL ApoE was quantitated by RIA. Plasma cholesterol and triglycerides were quantitated using an enzymatic assay. HDL cholesterol was determined in plasma following dextran sulfate precipitation [11]. For the initial data analysis the FCR of apoB-100 and apoE were calculated by curve fitting of the decaying part of the tracer curves, utilizing the CONSAM/SAAM 30 computer program. The tail of the amino acid tracer curve was subtracted to correct for the tracer reutilization. The residence time (RT) equals 1/FCR. At steady state FCR equals the fractional synthetic rate (FSR). The production rate (PR) of the protein is equal to the product of the (FCR) X (plasma concentration of the protein) X (plasma volume) divided by the body weight [3, 13]. Plasma volume was assumed to be 4% of body weight.

Results

The amino acids plasma curves were comparable between all four study subjects and did not show any differences between the male and female study subjects (data not shown). The VLDL apoB-100 curves of the two female subjects were comparable to each anoother and could be fit by almost identical curves, approaching both data points of each study subject equally close (Fig. 1). The same was true for the two male subjects (Fig. 2). Comparing the tracer uptake curves illustrates an earlier and higher peak for apoB-100 in the female subjects. The earlier peak indicates an increased FCR, whereas the peak height reflects the smaller pool seize. The FCR of VLDL apoB-100 is clearly increased for both female subjects in comparison to the two male subjects. Thus explains the lower VLDL apoB-100 levels in females due to an increased FCR. The kinetic data for VLDL apoB-100 are summarized in table 2.

Fig. 1 illustrates the tracer uptake curve into VLDL apoB-100 of the two female study subjects.

Fig. 2 illustrates the tracer uptake curve into VLDL apoB-100 of the two male study subjects.

Table 2. Initial kinetic data of VLDL apoB-100 metabolism

Subject	FCR (l/d)	RT (days)	conc (mg/dl)	PR (mg/kg-d)
1 (f)	10.09	0.099	2.7	10.8
2 (f)	10.00	0.100	2.8	11.0
3 (m)	7.67	0.130	3.6	11.1
4 (m)	7.67	0.130	5.3	16.4

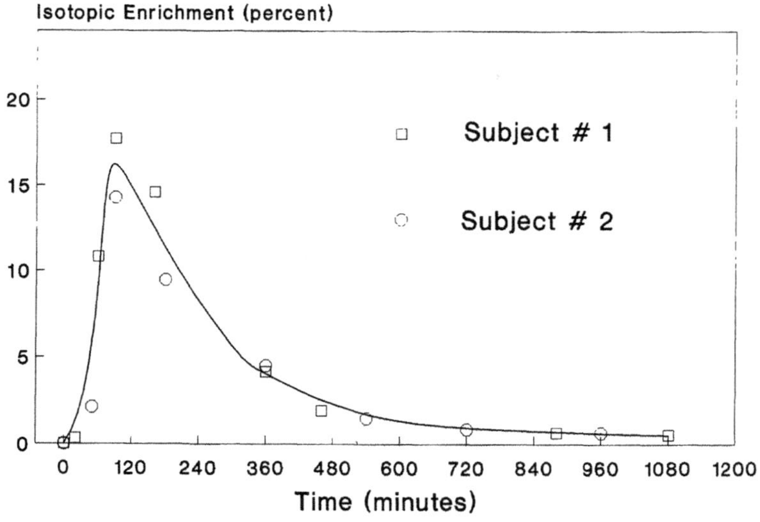

Fig. 1. VLDL ApoB-100 in the two female study subjects

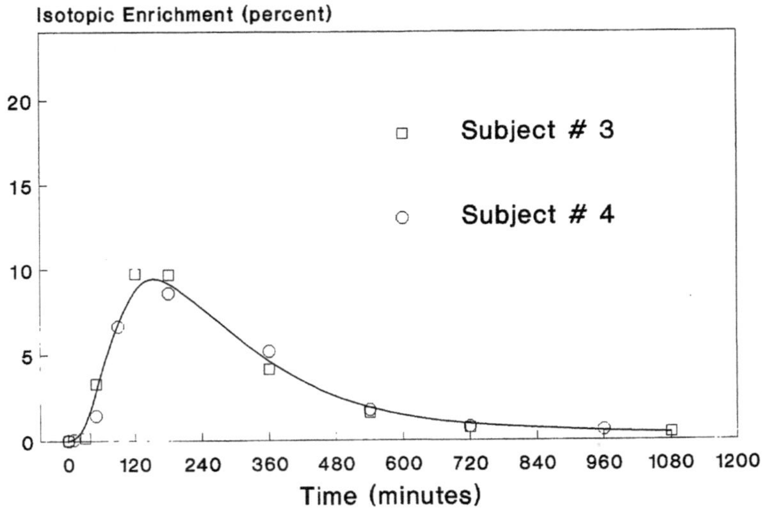

Fig. 2. VLDL ApoB-100 in the two male study subjects

The VLDL apoE curves did not show any major differences between the four study subjects and did not allow to perform data fitting for the study subjects corresponding to their sex. The tracer uptake curves are illustrated in figure 3 and the results are summarized in table 3. In this initial study we could not demonstrate sex related kinetic differences of the apoE metabolism. The PR and FCR of VLDL apoE was comparable for all four study subjects. Fig. 3 illustrates the tracer uptake curve into VLDL apoE of all study subjects.

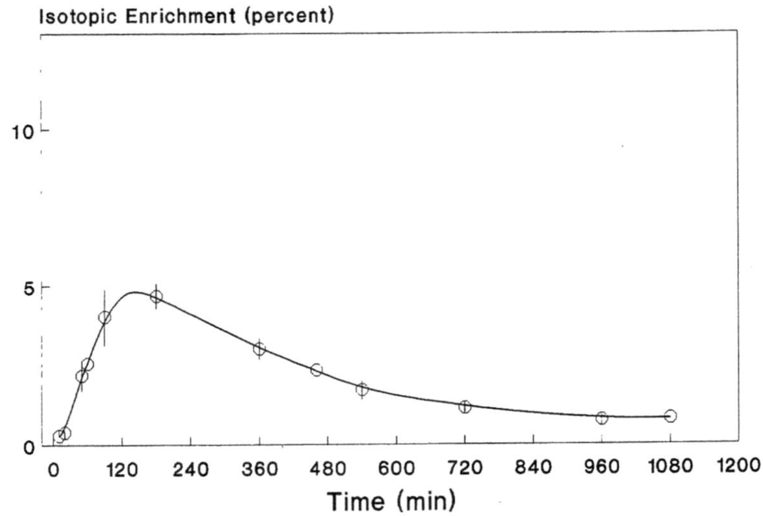

Fig. 3. VLDL ApoE in all study subjects

Table 3. Initial kinetic data of VLDL ApoE 3/3 metabolism

Subject	FCR (l/d)	RT (days)	conc (mg/dl)	PR (mg/kg-d)
1 (f)	3.3	0.30	1.2	1.6
2 (f)	3.6	0.28	1.0	1.4
3 (m)	4.0	0.25	1.0	1.6
4 (m)	3.0	0.33	1.4	1.7

Discussion

The present study describes a new technique for the in vivo investigation of proteins in humans. In vivo kinetic studies have traditionally been performed using radiotracer techniques [1, 8]. The use of radiotracers, however, has several major disadvantages. First, it exposes subjects to radioactivity and thereby excludes potential study subjects, such as children and pregnant women. Second, requires the isolation and exogenous labeling of proteins and thus allows the possibility that the protein may be modified by the isolation and labeling process or that it may be introduced into the study subject in a non-physiologic manner. Third, exogenous labeling with radioiodine limits the investigation to the simultaneous study of only two proteins due to the availability of only two usable iodine radioisotopes.

The stable isotope method described here overcomes all of these disadvantages. There is no use of ionizing radiation, thus expanding the pool of potential study subjects. The protein labeling is endogenous and physiologic, eliminating any possibility of alteration by isolation or iodination. Finally, the tracer is incorporated into all proteins being synthesized at the time of administration. This affords the opportunity to study virtually any protein, both within the plasma as well as potentially intracellular, that can be isolated. Thus the metabolism of multiple proteins can be simultaneously analyzed, a significant advantage over the dual label exogenous radiotracer approach.

These studies on sex related apolipoprotein in vivo kinetics indicate that the cause of lower triglyceride and VLDL apoB-100 levels in females is due to an increased fractional catabolic rate rather than differences on the production side. This is in agreement with earlier findings by Olefsky et al. who found that women clear very low density lipoprotein triglycerides more efficiently than men [7]. It is well known that estrogen can increase the apoB/E receptor activity, thus accelerating the catabolic side of apoB containing particles. The induction of the hepatic uptake of lipoproteins by estrogen is discussed in more detail by Windler [12]. The increase of VLDL binding to liver membranes by estrogen might explain in part the faster catabolism of apoB-100 in the two female study subjects compared to the males. On the otherhand Lackner and März [4] didn't find any difference in the apoB mRNA levels in cell culture studies under the influence of anabolic steroids. This observation is also in agreement with unchanged apoB production rates between the female and male study subjects in this initial report. The situation in respect of the apoE metabolism however is more complex. As apoE is exchangeable be-tween the different density classes the lack of kinetic differences of VLDL apoE in this primary study might be due to other apoE pools, such as HDL apoE. To elucidate precisely the apoE metabolism a more complex modeling is necessary, allowing to calculate the metabolic parameters within the distinct density classes. This report demonstrates that the stable isotope technique is a very useful tool for the *in vivo* investigation of protein metabolism in human.

Acknowledgements

We wish to thank the Pharmaceutical Development Branch, the NIH Clinical Center nurses from 8 East and the dietitian staff for their support, Yoshiko Doherty for technical support, Joanie Gault for preparation of the manuscript and especially the normal volunteers for participating in these studies.

J.R.S. was supported in part by a grant from the Deutsche Forschungsgemeinschaft (SCHA410/1-1) and the Kempkes Stiftung, University of Marburg, Federal Republic of Germany.

References

1. Fisher WR, Zech LA, Bardalaye P, Warmke G, Berman M (1980) The metabolism of apolipoprotein B in subjects with hypertriglyceridemia and polydisperse LDL. J Lipid Res 21:760–774.
2. Havel RJ, Eder HA, Bragdon JH (1955) The distribution and chemical composition of ultracentrifugally separated lipoproteins in human serum. J. Clin. Invest. 34:1345–1353.
3. Kensaniemi YA, Vega GL, Grundy SM (1982) Kinetics of apolipoprotein B in normal and hyperlipidemic man: Review of current data. In Lipoprotein Kinetics and Modeling. Berman M, Grundy SM, Howards BV, editors. Academic Press, New York, NY 181–205.
4. Lackner KJ, März W, Wolter G, Sartor H, Groß W; Anabolic steroids do not change m-RNA levels and protein secretion of apoA-I and apoB-100 in Hep-G2 cells; in this publication: 151–156.
5. MacKenzie SL, Tenaschuk D (1974) Gas-liquid chromatography of N-heptafluorobutyryl isobutyl esters of amino acids. J Chromatogr 97:19–24.
6. Meng MS, Gregg RE, Schaefer EJ, Hoeg JM, Brewer HB Jr (1983) Presence of two forms of apolipoprotein B in patients with dyslipoproteinemia. J. Lipid Res. 24:803–809.
7. Olefsky J, Farquhar JW, Reaven GM (1974) Sex difference in the kinetics of triglyceride metabolism in normal and hypertriglyceridaemic human subjects. Eur J Clin Invest 4:121–127.
8. Schaefer EJ, Zech LA, Jenkins LL, Bronzert TJ, Rubalcalba EA, Lindgren FT, Aamodt RL, Brewer HB Jr (1982) Human apolipoprotein A-I and A-II metabolism. J Lipid Res 23:850–862.
9. Schaefer JR, Rader DJ, Gregg RE, Fairwell T, Zech LA, Kindt MR, Benson MD, Brewer HB Jr (1990) In Vivo protein metabolism utilizing stable isotopes and mass spectrometry: a new approach to the study of mutant proteins in humans. Transactions of the Association of American Physicians CIII:187–194.

10. The Lipid Research Clinics Program Epidemiology Committee (1979) Plasma lipid distribution in selected North American populations: The Lipid Research Clinics Program Prevalence Study. Circulation 60:427–433.
11. Warnick GR, Benderson J, Albers JJ, Baillie EE, Sexton B, Schaefer EJ, Carlson D, Hill M, Brewer HB Jr, Wiebe DA, Hazlehurst J, Cooper GR (1982) Dextran sulfate-magnesium precipitation procedure for quantitation of high density lipoprotein cholesterol. Clin Chem 28:1379–1388.
12. Windler E, Jäckle S: Induction of the hepatic uptake of lipoproteins by estrogen. In this publication: 99–105.
13. Wolfe RR (1984) Whole body protein turnover. In Tracers in Metabolic Research. Radioisotope and Stable Isotope / Mass Spectrometry Methods. Wolfe RR, editor. Alan R Liss, Inc, New York, NY 157–173.
14. Wolfe RR (1984) Gas chromatography – mass spectrometry methods. In Tracers in Metabolic Research. Radioisotope and Stable Isotope / Mass Spectrometry Methods. Wolfe RR, editor. Alan R Liss, New York, NY 261–271.

Testosterone Lowers Serum Concentrations of Lipoprotein Lp(a)

M. W. BAUMSTARK, TH. v. STEIN, E. JAKOB, C. LULEY, A. BERG, and J. KEUL

Abstract

In 23 young males Lp(a) levels before and after application of testosterone have been studied. Testosterone (Testoviron Depot 250) was given at a dose of 2 × 250 mg with a 10 day interval (n = 14) or 6 × 250 mg with 7 day intervals. Lp(a) was measured by radioimmunoassay (Pharmacia-LKB, Freiburg). The test was standardized using a commercially available standard (Immuno, Heidelberg). Statistical comparisons were done by the non parametric Wilcoxon test. The testosterone treatment resulted in a slight elevation of LDL-cholesterol (+5%) as well as significant reductions of HDL (–10%, $p = 0.013$) and Lp(a) (–18%, $p = 0.008$) levels. Remarkably the change in Lp(a) was larger than changes in HDL or LDL. The mean Lp(a) concentration (n = 19) decreased from 30.2 to 24.7 mg/dl, the median from 13.5 to 11.1 mg/dl. In four subjects no Lp(a) could be detected (<2 mg/dl). These data confirm that Lp(a) and LDL levels are independently regulated and accordingly have to be modulated pharmacologically by different principles.

Introduction

The influence of different hormones on classical lipoprotein parameters has been studied by several groups (Review: [1]). However, only limited data exist on the influence of hormones on Lp(a) concentrations. In one paper Albers et al. [2] showed that in ten normolipidemic women with postmenopausal osteoporosis Lp(a) was reduced from 5.5 to 1.4 mg/dl by treatment with the anabolic steroid stanozolol (3 × 2 mg/d, 6 weeks). Harkes et al. [3] showed that estrogen treatment resulted in a 17-fold increase of the interaction of LDL with rat parenchymal liver cells whereas only a 2-fold effect on Lp(a) was found. Laplaud et al. [4] showed that Lp(a), like other lipoproteins, undergoes seasonal variations related to variations in testosterone levels. In this study we had the occasion to study lipoprotein parameters including Lp(a) in 23 males before and after testosterone administration.

Subjects and Methods

In 23 young clinically healthy males (23±7 J., 71±7 kg, 175±5 cm) Lp(a), lipid and lipoprotein levels before and after application of testosterone have been studied. Testosterone (Testoviron Depot 250) was given at a dose of 2×250mg with a 10 day interval (n=14) or 6×250mg with 7 day intervals (n=9). Since there were no statistically significant differences between these two groups, the data of both groups were pooled for the final evaluation.

Cholesterol and triglycerides were measured by enzymatic tests (Boehringer Mannheim); VLDL, LDL, and HDL cholesterol by quantitative lipoprotein electrophoresis (Immuno, Heidelberg). Lp(a) was measured by radioimmunoassay (Pharmacia-LKB, Freiburg). The test was standardized in mg/dl using a commercially available standard (Immuno, Heidelberg). Statistical comparisons were done by the non parametric Wilcoxon test for paired samples.

Table 1. Lipoprotein parameters before and after testosterone administration (n=19)

	before	after	p
Chol	189±38	192±14	n.s.
TG	109±56	137±59	n.s.
VLDL	17± 8	20± 9	n.s.
LDL	114±26	120±13	n.s.
HDL	58±13	52±13	0.013
Lp(a)	30±31	24±26	0.008

mean ± SD. Units: mg/dl

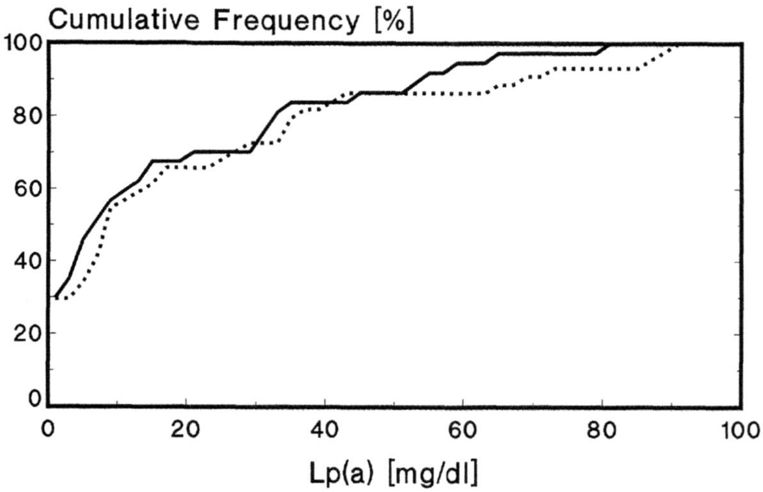

Fig. 1. Cumulative frequency distribution of Lp(a) concentrations before (......) and after (——) administration of testosterone (n=23)

Results

As shown in Table 1 the testosterone treatment resulted in slight but not significant elevations of cholesterol (+2%), triglycerides (+26%), LDL-cholesterol (+5%), and VLDL-cholesterol (+18%). HDL was reduced significantly by 10% ($p=0.013$). Even more significant was the reduction of Lp(a) (–18%, $p=0.008$). Remarkably, the change in Lp(a) was larger than changes in HDL or LDL. The mean Lp(a) concentration ($n=19$) decreased from 30.2 to 24.7 mg/dl, the median from 13.5 to 11.1 mg/dl. It should be noted that in four subjects Lp(a) levels were below 2 mg/dl at both times of measurement. Figure 1 shows the cumulative frequency of Lp(a) concentrations before and after treatment. The largest decrease of about 30% was measured in two individuals with Lp(a) levels above 80 mg/dl.

Discussion

The changes of the lipoprotein profile demonstrated in this study are consistent with the established mechanisms of action [1, 5, 6] of androgenic steroids: Decreased HDL levels are primarily due to an increased activity of hepatic lipase [5, 6]. In addition a reduced synthesis rate of VLDL, ApoA-I, and ApoA-II has also been discussed [5] as contributing factor for decreased HDL concentrations. If testosterone has any impact on the LDL receptor activity at all, it decreases the activity of this receptor, which could explain the increase of LDL and total cholesterol in this study. In contrast to slightly higher LDL concentrations we found significantly decreased Lp(a) levels during testosterone administration. The fact that testosterone lowers Lp(a) but rises LDL shows that the metabolism of both particles is regulated differently. This is according to data on HMG-CoA reductase inhibitors [7, 8], which show that an increase in LDL receptor activity lowers LDL but not Lp(a). In contrast, overexpression of human LDL receptors leads to accelerated catabolism of Lp(a) in transgenic mice [9].

Hypothetical reasons for reduction of Lp(a) under androgenic steroids might be a reduction of hepatic synthesis of apo(a) or an accelerated catabolism of Lp(a) by the upregulated hepatic lipase. The present data confirms results [2] demonstrating that Lp(a) concentrations can be lowered by androgenic steroids. However, in the study of Albers stanozolol, which is known to have a very strong influence on lipid metabolism, was given to postmenopausal women, while in this study testosterone was given to men. It should be noted hat nearly all known lipid lowering interventions including diet [10] and physical exercise [11] are not able to lower Lp(a) concentrations. The only other substances, except anabolic steroids, that have been reported to reduce Lp(a) are drugs [12, 13] or drug combinations [14] containing nicotinic acid. Recently there have been reports on lowering of Lp(a) with n-acetylcysteine [15, 16] or sustained-release bezafibrate [17], findings, however, that have to be confirmed independently.

References

1. Godsland IF, Wynn V, Crook D, Miller NE (1987) Sex, plasma lipoproteins, and atherosclerosis: prevailing assumptions and outstanding questions. Am Heart J 114:1467–1502.
2. Albers JJ, Taggart HM, Applebaum-Bowden D, Haffner S, Chesnut CH, Hazzard WR (1984) Reduction of lecithin-cholesterol acyltransferase, apolipoprotein D and the Lp(a) lipoprotein with the anabolic steroid stanozolol. Biochim Biophys Acta 795:293–296.
3. Harkes L, Jürgens G, Holasek A, Van Berkel TJ (1988) In vivo studies on the binding sites for lipoprotein (a) on parenchymal and non-parenchymal rat liver cells. FEBS Lett 227:27–31.
4. Laplaud PM, Saboureau M, Beaubatie L, el Omari B (1989) Seasonal variations of plasma lipids and lipoproteins in the hedgehog, an animal model for lipoprotein (a) metabolism: relation to plasma thyroxine and testosterone levels. Biochim Biophys Acta 1005:143–156.
5. Haffner SM, Kushwaha RS, Foster DM, Applebaum-Bowden D, Hazzard WR (1983) Studies on the metabolic mechanism of reduced high density lipoproteins during anabolic steroid therapy. Metabolism 32:413–420.
6. Applebaum-Bowden D, Haffner SM, Hazzard WR (1987) The dyslipoproteinemia of anabolic steroid therapy: increase in hepatic triglyceride lipase precedes the decrease in high density lipoprotein2 cholesterol. Metabolism 36:949–952.
7. Leren TP, Hjermann I, Berg K, Leren P, Foss OP, Viksmoen L (1988) Effects of lovastatin alone and in combination with cholestyramine on serum lipids and apolipoproteins in heterozygotes for familial hypercholesterolemia. Atherosclerosis 73:135–141.
8. Thiery J, Armstrong VW, Schleef J, Creutzfeldt C, Creutzfeldt W, Seidel D (1988) Serum lipoprotein Lp(a) concentrations are not influenced by an HMG CoA reductase inhibitor. Klin Wochenschr 66:462–463.
9. Hofmann SL, Eaton DL, Brown MS, Mcconathy WJ, Goldstein JL, Hammer RE (1990ß) Overexpression of human low density lipoprotein receptors leads to accelerated catabolism of lp(a) lipoprotein in transgenic mice. J Clin Invest 85: 1542–1547.
10. Masarei JR, Rouse IL, Lynch WJ, Robertson K, Vandongen R, Beilin LJ (1984) Effects of a lacto-ovo vegetarian diet on serum concentrations of cholesterol, triglyceride, HDL-C, HDL2-C, HDL3-C, apoprotein-B, and Lp(a). Am J Clin Nutr 40:468–478.
11. Stein T, Baumstark MW, Berg A, Keul J (1991) Serum concentration of lipoprotein Lp(a) in athletes and sedentary men. In: Bernett P, Jeschke D (eds) Sport und Medizin, Pro und Contra. Zuckschwerdt-Verlag, München, pp 530–532.
12. Carlson LA, Hamsten A, Asplund A (1989) Pronounced lowering of serum levels of lipoprotein Lp(a) in hyperlipidaemic subjects treated with nicotinic acid. J Intern Med 226:271–276.
13. Noma A, Maeda S, Okuno M, Abe A, Muto Y (1990) Reduction of Serum Lipoprotein(a) Levels in Hyperlipidaemic Patients with Alpha-Tocopheryl Nicotinate. Atherosclerosis 84:213–217.
14. Gurakar A, Hoeg JM, Kostner G, Papadopoulos NM, Brewer B (1985) Levels of lipoprotein Lp(a) decline with Neomycin and Niacin treatment. Arteriosclerosis 57:293–301.
15. Gavish D, Breslow JL (1991) Lipoprotein(a) Reduction by N-Acetylcysteine. Lancet 337:203–204.
16. Scanu AM (1991) N-Acetylcysteine and Immunoreactivity of Lipoprotein(a). Lancet 1991; 337:1159.
17. Bimmermann A, Boerschmann C, Schwartzkopff W, Vonbaeyer H, Schleicher J (1991) Effective Therapeutic Measures for Reducing Lipoprotein (a) in Patients with Dyslipidemia – Lipoprotein (a) Reduction with Sustained-Release Bezafibrate. Curr Ther Res 49:635–643.

The Role and Mechanism of Glucocorticoids on the Modulation of Lipid and Lipoprotein Metabolism

R. E. GREGG

Introduction

Glucocorticoid hormones have significant effects on the metabolism of lipids and lipoproteins. This review will briefly summarize the mechanism of action of glucocorticoid hormones, and then will describe some of the abnormalities in lipid and lipoprotein metabolism secondary to alterations in glucocorticoid levels that result in altered circulating concentrations of lipids and lipoproteins.

Glucocorticoids are a class of steroid derived hormones that exert their effects through interactions with glucocorticoid receptor proteins [1]. These hormone receptors are part of the steroid hormone receptor superfamily which includes other steroid hormone receptors, thyroid hormone receptors, retinoid hormone receptors, and a number of orphan receptors without known ligands. These receptors are made up of a number of functional domains including a ligand binding domain, a DNA binding domain, a dimerization domain, one or more transactivation domains, a nuclear localization domain, and the glucocorticoid hormone receptor has an hsp90 binding domain.

The glucocorticoid receptor is present in the cytoplasm of hormone responsive cells complexed with the 90 kDa heat shock protein (hsp90) in the absence of hormone [1]. When the receptor interacts with its ligand, it dissociates from hsp90, forms a dimer, and this activated receptor dimer is transported into the nucleus and binds to specific DNA regulatory elements, called glucocorticoid response elements (GRE's), which results in transcriptional regulation of the associated gene. The diversity of the regulatory actions of steroid hormones results from the modulation of the hormone interactions with receptors and hormone response elements at a number of different levels. These include tissue specific modification of hormone structure, tissue specific expression of hormone receptors, interactions with hsp90, homodimerization or heterodimerization with other steroid hormone receptors, competition for binding with other nuclear factors (squelching), and interactions with the GRE's. The GRE's provide additional diversity in the specificity of action of glucocorticoids resulting from minor changes in the sequences of the GRE's, the number of tandem copies of the GRE in the regulatory elements of a gene, the position of the GRE('s) in the regulatory elements, and the interactions of the receptors with other cell-specific and gene-specific regulatory factors present in the nucleus.

Mechanism of Glucocorticoid Effects on Lipid and Lipoprotein Metabolism

All of the know actions of glucocorticoids are generated either directly or indirectly through the binding with the glucocorticoid hormone receptor resulting in transcriptional regulation of gene expression. There are a number of genes that are known to contain glucocorticoid response elements whose transcription is regulated by glucocorticoids. However, there are no convincing data at this time to indicate that either apolipoprotein genes or lipoprotein receptor genes are directly regulated through GRE/glucocorticoid hormone receptor interactions. Thus, the effects of glucocorticoids on lipoprotein metabolism are communicated indirectly through the modulation of other hormones, enzymes, and receptors.

Increased circulating levels of glucocorticoids, either from an endogenous or an exogenous origin, effect the levels of multiple other hormones. They will increase the secretion of insulin (possibly secondarily to glucocorticoid induced insulin resistance in peripheral tissues) and glucagon from pancreatic islets [2, 3], increase the secretion of growth hormone, increase the secretion of epinephrine from the adrenal medulla [3], and inhibit the conversion of T_4 into T_3. These alterations by glucocorticoids have the end result of inducing insulin resistance (elevations in levels of glucagon, growth hormone, and epinephrine) and down regulating the LDL receptor (inhibition of T_3 formation) which results in elevations in circulating lipid and lipoprotein levels.

In addition to the effects on the levels of other hormones, elevated levels of glucocorticoids also result in a number of metabolic effects that modulate lipid and lipoprotein metabolism. These include an induction in the resistance to the actions of insulin in peripheral tissues [2, 3, 4]. In a study by Taskinen, et al. in subjects with Cushing's syndrome [5], the fasting glucose was only 10% greater than the control subjects, but one and two hours following the ingestion of a standard glucose load, the Cushing's patients had glucose levels that were more than 65% greater than the controls. Their fasting insulin levels were nearly double the controls, and this difference was maintained during the oral glucose tolerance test. These results are characteristic of the resistance to the actions of insulin found in subjects with excess glucocorticoids.

Rizza, et al., demonstrated that in normal subjects infused with cortisol for 24 hours, there was a shift to the right in the insulin dose response curves for both suppression of glucose production and stimulation of glucose utilization [4]. They concluded from these and additional studies, that cortisol induced an insulin resistance in both the liver and peripheral tissues, and that this resistance was due to a postreceptor defect in insulin action.

Glucocorticoid excess can result in an inhibition of intravascular lipolysis of triglycerides [6] which can result in hypertriglyceridemia. In addition, high levels of glucocorticoids can lead to the stimulation of intracellular lipolysis either directly [7] or secondarily to glucocorticoid induced insulin resistance. Increased levels of circulating free fatty acids are transported to the liver where they are either metabolized further or converted into triglycerides and secreted as VLDL.

An increased caloric intake can be induced with the administration of glucocorticoids [8] resulting in an increase in carbohydrate and fat intake. There is also an increase in the catabolism of cellular protein with the excess two and three carbon fragments being shunted into glucose, fatty acids, and triglycerides with a decrease in ketone body formation [9]. The increase in free fatty acid flux into the liver, the increased caloric intake, and the catabolism of proteins and amino acids all lead to an increase in the substrates necessary for increased hepatic triglyceride synthesis and VLDL secretion.

Glucocorticoids can also cause a down regulation in the activity of LDL receptors. By inhibition of the conversion of T_3 to T_4, they can decrease the transcription of the LDL receptor. In addition, glucocorticoids can inhibit LDL receptor dependent internalization and degradation of LDL in cultured skin fibroblasts and aortic endothelial cells [10] while having little effect on LDL binding to its receptor. These abnormalities can result in an elevation in the levels of circulating cholesterol and apolipoprotein B.

Of interest, individuals with glucocorticoid excess have been demonstrated to have low levels of hepatic lipase activity [5]. A low level of hepatic lipase has been previously shown to be associated with elevated HDL cholesterol levels, as is found in individuals with glucocorticoid excess.

These combined metabolic effects result in insulin resistance with an elevation in plasma glucose levels. Additional alterations in intermediary metabolism provide the substrates necessary for increased triglyceride production with increased secretion of chylomicrons and VLDL. The defects in LDL receptor function can result in a decreased fractional catabolic rate for LDL receptor dependent lipoprotein catabolism, and the decreased hepatic lipase activity has been associated with elevations in HDL levels.

Physiological Effects of Glucocorticoids on Lipids and Lipoproteins

What are the actual alterations in glucose, lipid, and lipoprotein levels found in subjects with glucocorticoid excess? As one would predict from the metabolic abnormalities found in these subjects, there are elevations in the levels of plasma glucose, cholesterol, triglyceride, chylomicrons, VLDL, LDL, and HDL. These elevations are variable and are not found in all individuals, but they are frequently seen in individuals with glucocorticoid excess. The most common abnormalities are insulin resistance with near normal fasting glucose levels and elevations in triglyceride and VLDL levels. Frequent, but less common, are elevations in fasting glucose levels, and elevations in cholesterol and LDL levels with or without increases in VLDL levels. One also frequently finds moderate elevations in HDL cholesterol levels. An uncommon finding is an extreme elevation in triglyceride levels with the chylomicronemia syndrome. In a study of eleven untreated subjects with Cushing's syndrome compared to eleven matched control subjects [5], the following elevations in plasma levels were found: glucose 10%, triglycerides 74%, cholesterol 24%, VLDL cholesterol 100%, LDL cholesterol 19%, and HDL cholesterol 21%.

The most consistent kinetic abnormality in lipoprotein metabolism secondary to excess glucocorticoids is an overproduction of VLDL. From freshly isolated rat hepatocytes, it has been demonstrated that there is an increase in VLDL secretion when these cells are incubated with dexamethasone in tissue culture [11, 12, 13]. When rats were treated with dexamethasone for seven days followed by liver perfusion studies, it was found that the treated rat livers secreted more total triglyceride and more VLDL triglyceride [9]. In kinetic studies on the above described eleven patients with Cushing's syndrome [5], it was determined that the elevated plasma triglyceride and VLDL levels were the result of a 40% increase in the production rate of VLDL triglyceride with little change in its fractional catabolic rate. Thus, the data are very consistent that there is an induction in the production rate of VLDL by glucocorticoids. There is also evidence that when high levels of exogenous glucocortoids are administered to individuals, heparin releasable lipoprotein lipase [6] and hepatic lipase [5] activities are decreased. Less well documented are in vivo confirmations of the evidence for decreased LDL receptor activity secondary to glucocorticoid administration [10].

There are multiple factors described above that can up regulate VLDL synthesis and secretion, but the data are most consistent with the increased VLDL production being secondary to a combination of the direct effects of glucocorticoids on intermediary metabolism and the metabolic effects secondary to the insulin resistance induced by this class of compounds. In the study by Taskinen, et al. [5], there was a statistically significant association between by the insulin response to an oral glucose challenge (an indirect measure of insulin resistance) and the VLDL triglyceride production rate. In addition to the direct effects of glucocorticoids, insulin resistance will result in an increase in plasma free fatty acid levels with increased uptake by the liver, increased hepatic uptake of glucose and lactate, increased conversion of substrates into triglyceride, and an increase in apoB and VLDL secretion [14, 15].

Conclusion

In summary, glucocorticoids have multiple effects on carbohydrate and lipid metabolism as well as on the modulation of hormones that effect these metabolic pathways. Excess glucocorticoids result in insulin resistance, hyperglycemia, and hypertriglyceridemia with elevated VLDL levels. Less consistently seen are elevations in LDL and HDL levels. The major cause of the elevation in lipids and lipoproteins is an over production of VLDL by the liver with contributory effects coming from a decrease in the clearance of lipids and lipoproteins from the circulation.

References

1. Muller M, Renkawitz R (1991) The glucocorticoid receptor. Biochim. et Biophys. Acta 1088:171–182.
2. Marco J, Calle C, Hedo JA, Villanueva ML (1976) Enhanced glucagon secretion by pancreatic islets from prednisolone-treated mice. Diabetologia 12:307–311.
3. Eigler N, Sacca L, Sherwin RS (1979) Synergistic interactions of physiologic increments of glucagon, epinephrine, and cortisol in the dog. J Clin Invest 63:114–123.
4. Rizza RA, Mandarino LJ, Gerich JE (1982) Cortisol-induced insulin resistance in man: impaired suppression of glucose production and stimulation of glucose utilization due to a postreceptor defect of insulin action. J Clin Endo Metab 54: 131–138.
5. Taskinen M-R, Nikkila EA, Pelkonen R, Sane T (1983) Plasma lipoproteins, lipolytic enzymes, and very low density lipoprotein triglyceride turnover in Cushing's syndrome. J Clin Endo Metab 57:619–626.
6. Bagdade JD, Porte D, Bierman EL (1970) Steroid-induced lipemia. Arch Intern Med 125:129–134.
7. Divakaran P, Friedmann N (1976) A fast in vitro effect of glucocorticoids on hepatic lipolysis. Endocrinology 98:1550–1553.
8. Krotkiewski M, Bjorntorp P (1975) The effects of dexamethasone and starvation on body composition and regional adipose tissue cellularity in the rat. Acta Endocrinologica 80:667–675.
9. Cole TG, Wilcox HG, Heimberg M (1982) Effects of adrenalectomy and dexamethasone on hepatic lipid metabolism. J Lipid Res 23:81–91.
10. Henze K, Chait A, Albers JJ, Bierman EL (1983) Hydrocortisone decreases the internalization of low density lipoprotein in cultred human fibroblasts and arterial smooth muscle cells. European J Clin Invest 13:171–177.
11. Brindley DN (1981) Regulation of hepatic triacylglycerol synthesis and lipoprotein metabolism by glucocorticoids. Clinical Science 61:129–133.
12. Mangiapane EH, Brindley DN (1986) Effects of dexamethasone and insulin on the synthesis of triacylglycerols and phosphatidylcholine and the secretion of very low-density lipoproteins and lysophosphatidylcholine by monolayer cultures of rat hepatocytes. Biochem J 233:151–160.
13. Bartlett SM, Gibbons GF (1988) Short- and longer-term regulation of very-low density lipoprotein secretion by insulin, dexamethasone and lipogenic substrates in cultured hepatocytes. Biochem J 249:37–43.
14. Tobey TA, Greenfield M, Kraemer F, Reaven GM (1981) Relationship between insulin resistance, insulin secretion, very low density lipoprotein kinetics, and plasma triglyceride levels in normotriglyceridemic man. Metabolism 30:165–171.
15. Howard BV (1987) Lipoprotien metabolism in diabetes mellitus. J Lipid Res 28: 613–628.

Lipids and Lipoproteins in Thyroid Dysfunction

J. DABELS and E. FINGER

There has been a longstanding interest in lipid metabolism and thyroid dysfunction since very early observations of high total cholesterol level in hypothyroidism by Mason et al. in 1930. A relationship between serum lipid concentrations and basal metabolic rate was already found by Epstein and Lande in 1922. The mechanisms responsible for the various dyslipoproteinemias are still incompletely unterstood in detail. On the other hand relationship between thyroid hormones and alterations of key enzymes of the lipoprotein metabolism are very likely to be the reason behind disturbances in lipoprotein metabolism in thyroid diseases [15, 20]. It should also be noted that most

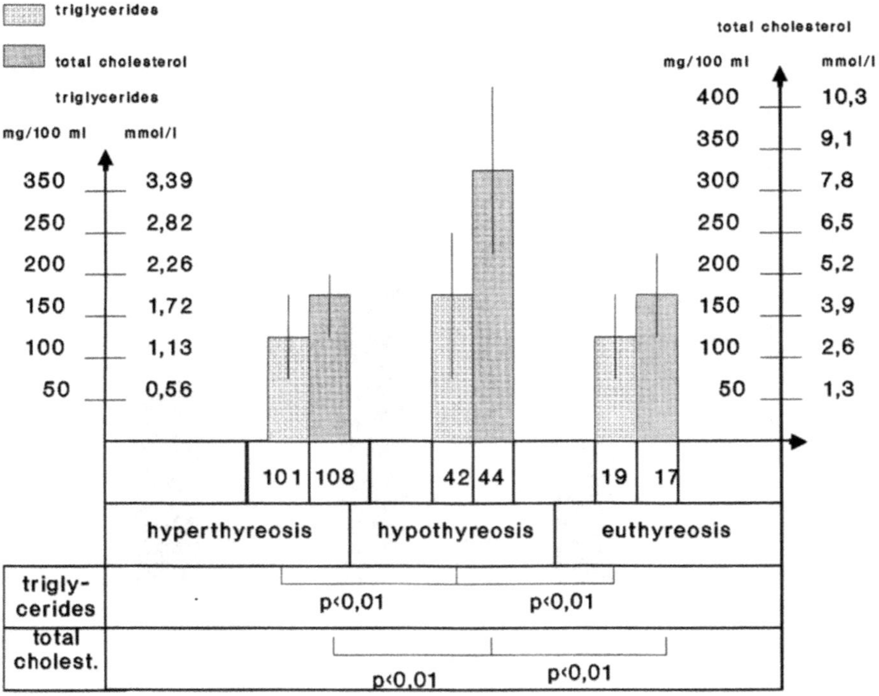

Fig. 1. Triglycerides and total cholesterol in hypo-, hyper- and euthyroid patients ($\bar{x} \pm s_x$)

probably the LDL receptor, the LDL internalization and/or the intracellular LDL metabolism may be influenced by thyroid hormones [7, 11, 18].

The alterations are most prominent in hypothyroidism which is typically accompanied by hypercholesterolemia and frequently but not always also by moderate hypertriglyceridemia [1, 6, 9, 19, 22]. Hyperthyroidism on the other hand is often associated with decreased levels of cholesterol [1]. These changes are completely reversibel by either thyroxin replacement therapy or by thyreostatic treatment. The same is published concerning concentration of apolipoprotein B, which was increased in hypothyroidism and normalized by replacement therapy [5, 11].

Our own studies in 119 hyperthyroid and 43 hypothyroid patients and 20 normal persons reveal elevated total as well as LDL-cholesterol levels in the serum of hypothyroids. Hyperthyroid patients in contrast showed reduced total and LDL-cholesterol levels. The same pattern was seen in triglycerides (Fig. 1 and 2). All patients were investigated after giving informed consent and after approval by the local ethic committee.

A relation between body weight and triglyceride levels can't be excluded. The differences between the mean values of free fatty acids were not significant (Fig. 3 and 4).

Total cholesterol and LDL cholesterol showed a negative correlation to the triiodothyronine values in hyperthyroid patients (Fig. 5, 6).

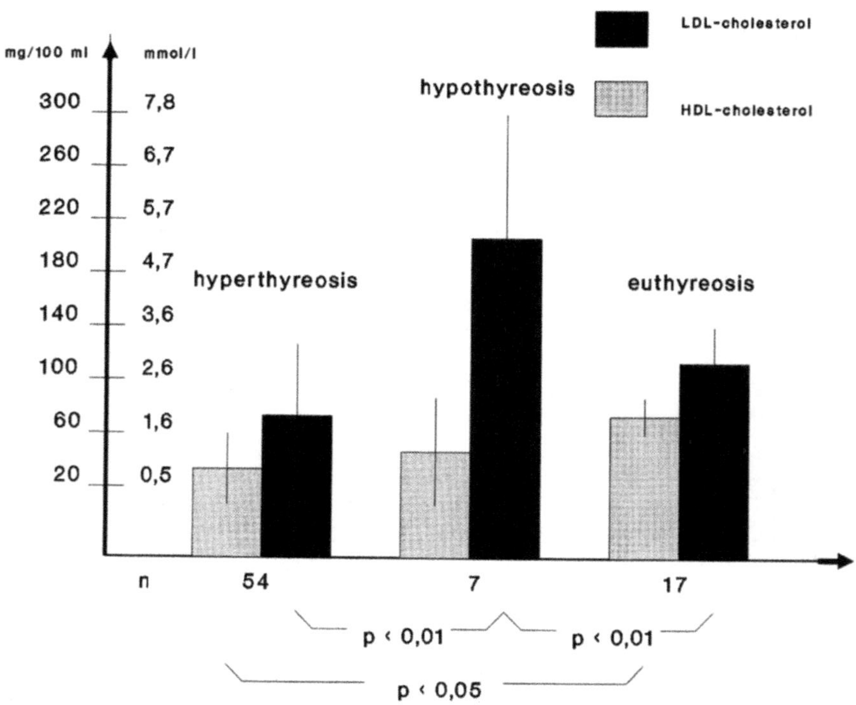

Fig. 2. LDL- and HDL-cholesterol in hypo-, hyper- and euthyroid patients ($\bar{x} \pm s_x$)

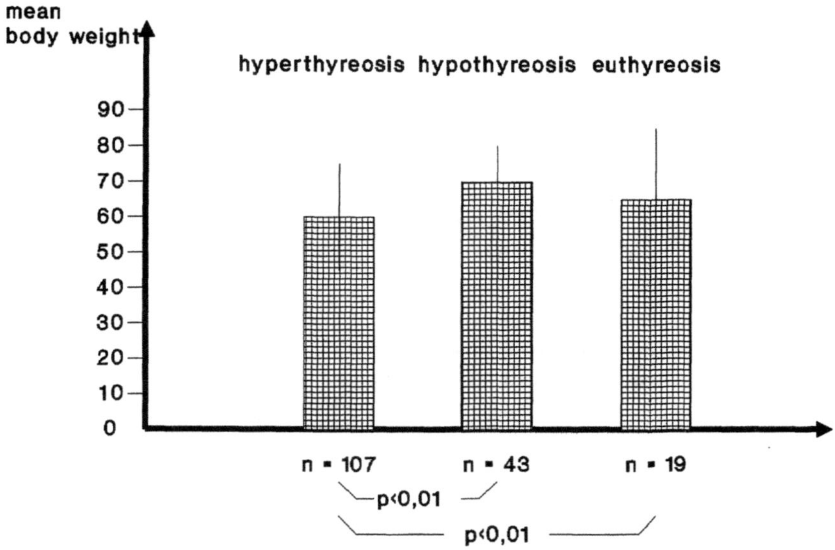

relative overweight	$\bar{x} \pm s_x$ (n)	$\bar{x} \pm s_x$ (n)	$\bar{x} \pm s_x$ (n)
	8,1±19,5(107)	23,4 ± 21,1 (42)	15,5 ± 29,3 (19)

p<0,01

Fig. 3. Relative overweight in hypo-, hyper- and euthyroid patients ($\bar{x} \pm s_x$)

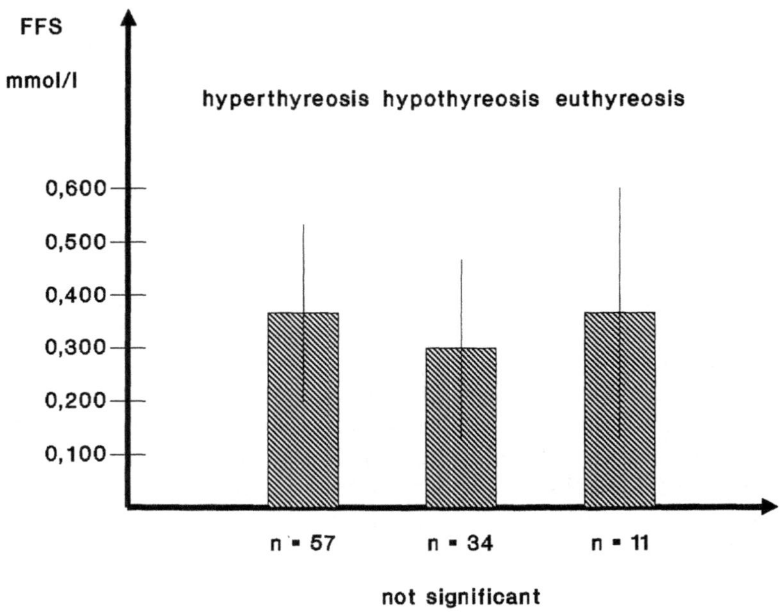

Fig. 4. Free fatty acids in hypo-, hyper- and euthyroid patients ($\bar{x} \pm s_x$)

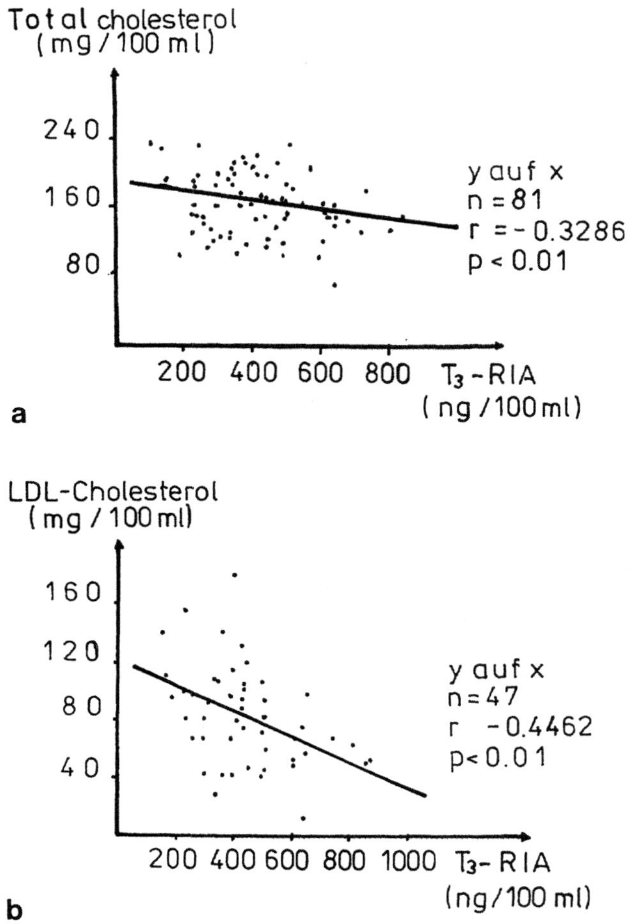

Fig. 5 a, b. Total cholesterol and LDL-cholesterol in relation to triiodothyronine levels in hyperthyroid patients

High density lipoproteins which are responsible for the transport of cholesterol from the peripheral organs to the liver were seen to be reduced in hypo- and hyperthyroidism either (Fig. 7). In literature conflicting data are reported in view of the HDL, in hypothyroidism increased [16], in hyperthyroidism reduced or unchanged [1].

Nevertheless increased LDL cholesterol levels in serum should be given more weight than HDL-levels regarding the risk of coronary heart disease in thyroid hormone deficient state [11]. Therefore it is more significant to calculate the LDL-HDL-ratio, which was found to be much higher in hypothyroid patients than in both other groups (Fig. 8).

Between patients with primary or secondary hypothyroidism there was no difference in the LDL-HDL-ratio [20].

To understand the mechanisms behind we had to focus on the key enzymes of lipoprotein metabolism that means on the postheparin lipolytic ac-

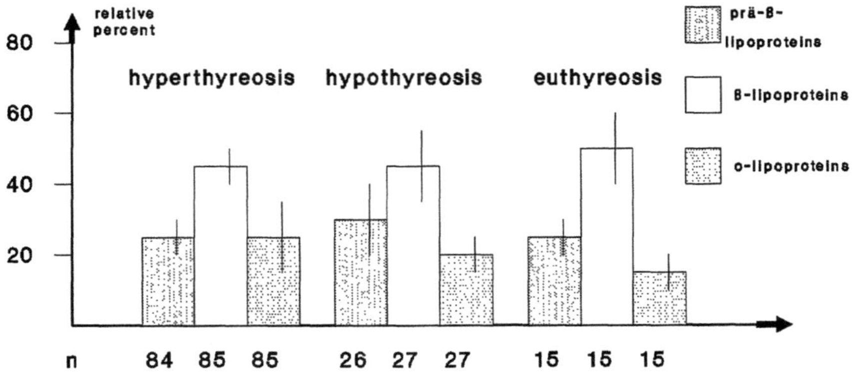

	hyperthyreosis	hypothyreosis	euthyreosis
Prä-ß-LP	not significant		
ß-LP	p < 0,05 (hyperthyreosis – euthyreosis); p < 0,05 (hypothyreosis – euthyreosis)		
α-LP	p < 0,01 (hyperthyreosis – hypothyreosis); p < 0,01 (hyperthyreosis – euthyreosis)		

Fig. 6. Lipoproteins in hypo-, hyper- and euthyreoisis

$$\frac{\text{LDL - Cholesterol}}{\text{HDL - Cholesterol}}$$

hyperthyreosis	hypothyreosis	euthyreosis
$\bar{X} \pm S_{\bar{x}}$ (n)	$\bar{X} \pm S_{\bar{x}}$ (n)	$\bar{X} \pm S_{\bar{x}}$ (n)
1,69 ± 0,9 (54)	4,01 ± 1,85 (7)	1,5 ± 0,76 (17)

p<0,01 (hyperthyreosis – hypothyreosis)

p<0,01 (hypothyreosis – euthyreosis)

Fig. 7. LDL/HDL-cholesterol ratio in hypo-, hyper- and euthyreosis

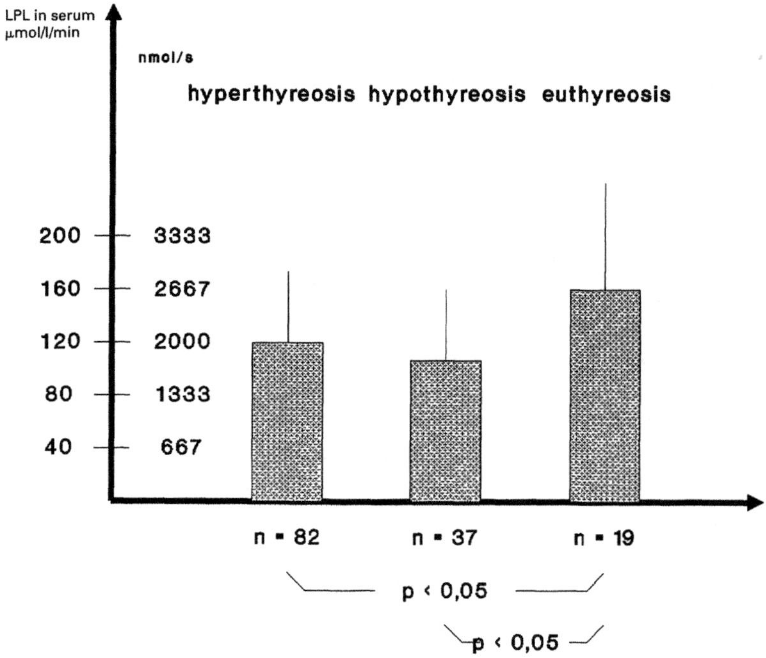

Fig. 8. Postheparin lipoprotein lipase in hypo-, hyper- and euthyreosis

tivity (PHLA) subdivided in hepatic lipase (HL) and lipoprotein lipase (LPL).

Our results are in accordance with those obtained previously by other authors who have found a reduced LPL activitiy expressed as FFA release from an Intralipid substrate in hypothyroidism [12, 14, 19]. In hyperthyroidism on the other side reports on LPL are contradictory [1, 8, 12, 19]. In our own study we found a moderately decreased LPL activity also in hyperthyroid patients. We found a positive correlation between the LPL in serum of hypothyroid patients and the effective thyroid ratio (ETR) (Fig. 9).

Differentiating PHLA in LPL and HL reveals normal (till) slowly reduced LPL levels but reduced HL in hypothyroidism, and increased HL but normal LPL in hyperthyroidism [1].

The reason of low LPL activities in hypothyroidism may be due to effects on the synthesis, posttranslational modifications on the secretion rate or on the glycosaminoglycan receptors of the endothelial cells [13, 14]. The metabolism of glycosaminoglycans is known to be changed in hypothyroidism [23]. A rise in the LPL activity may be due to a non specific effect of triiodothyronine [17]. The effect of thyroid hormones on the HL is more direct and is followed by an effect on the HDL_2-fraction. This particle is product of LPL and also substrate of HL either [20, 21].

In the ultracentrifuged plasma of hyperthyroid patients a special subfraction with β-mobility was shown [24], which couldn't be separated from the HDL-fraction. The protein-lipid relation in this lipoprotein shifted from 22 to 43% of protein whilst the degree of cholesterol decreased from 45 to 25 % compared with LDL particles. This so-called β-HDL fraction has been found in 39 of 46 hyperthyroid patients and disappeared without exception after treatment.

Summarizing our results we can draw the conclusion that thyroid hormones produce profound alterations in lipid and lipoprotein metabolism which have to be considered in pathochemistry and treatment of thyroid dysfunction state.

References

1. Abrams JJ, Grundy SM (1981) Cholesterol metabolism in hypothyroidism and hyperthyroidism in man. J Lipid Res 22:323–338.
2. Agdeppa D, Macaron C, Mallik T, Schnuda ND (1979) Plasma high density lipoprotein cholesterol in thyroid disease. J Clin Endocrinol Metab 49:726–729.
3. Arons DL, Schreibman PH, Downs P, Braverman LE, Arky RA (1972) Decreased post heparin lipases in Grave's disease. N Engl J Med 286:233–237.
4. Epstein AA, Lande H (1922) Studies on blood lipoids I. The rela tion of cholesterol and protein deficiency to basal meta bolism. Arch Int Med 30:563–577.
5. Friis T, Pedersen LR (1987) Serum lipids in hyper- and hypo thyroidism before and after treatment. Clin Chim Acta 162:155.
6. Hazzard WR, Bierman EL (1972) Aggravation of broad B-disease in hypothyroidism. Arch Int Med 130:822–825.
7. Hoch FL (1988) Lipids and thyroid hormones. Prog Lipid Res 27:199–270.
8. Kirkeby K (1968) Postheparin plasma lipoprotein lipase activity in thyroid disease. Acta Endocrinol (Kbh) 59:555–563.
9. Kutty KM, Bryant DG, Farid NR (1978) Serum lipids in hypothyroidism 46: 55–60.
10. Mason RL, Hunt HM, Hurxthal L (1930) Blood cholesterol values in hyperthyroidism and hypothyroidism – their significance. N Engl J Med 203:1273–1278.
11. Muller MJ, Seitz HJ (1984) Thyroid hormone action on intermediary metabolism. Part II: Lipid Metabolism in hypo- and hyperthyroidism. Klin Wschr 62:49–55.
12. Nikkila EA, Kekki M (1972) Plasma triglyceride metabolism in thyroid disease. J Clin Invest 51:2103–2114.
13. Nilsson-Ehle P Regulation of lipoprotein lipase. In: Carlsson LA & Pernow B (eds). Metabolic risk factors in ischemic cardiovascular disease. 35–48, Raven Press New York.
14. Pykalisto O, Goldberg AP, Brunzell JD (1976) Reversal of decreased human adipose tissue lipoprotein lipase and hypertriglyceridemia after treatment of hypothyroidism. J Clin Endocrinol Metab 43:591–600.
15. Rassoul F, Richter V, Kuhlmann E, Heberling HJ, Homann M, Zuck U, Rotzsch W, Lohmann D (1988) Lipoproteinstoffwechsel bei Patienten mit gestörter Schilddrüsenfunktion vor und nach Therapie. Dtsch Z Verdau-Stoffwechskrankh 48:104–111.
16. Scottolini AG, Bhagavan NV, Oshiro TH (1980) Serum high-density lipoprotein cholesterol concentrations in hypo- and hyper thyroidism. Clin Chem 26:584–587.
17. Tata JR, Widnell CC (1966) Ribonucleic acid synthesis during the early action of thyroid hormones. Biochem J 98:604–620.
18. Thompson GR, Soutar, AK, Spengel FA, Jadhar A, Gavigan SJP, Myant NB (1981) Defects of receptor-mediated low density lipo protein catabolism in homozygous familial hypercholesterol emia and hypothyroidism in vivo. Proc Natl Acad Sci USA 78:2591–2595.

19. Tulloch, BR, Lewis B, Russell Frazer T (1973) Triglyceride metabolism in thyroid disease. Lancet I:391–394.
20. Valdemarsson St (1983) Plasma lipoprotein alterations in thyroid dysfunction. Roles of lipoprotein lipase, hepatic lipase and LCAT. Doctoral Dissertation. Lund, p 9–13.
21. Valdemarsson S, Hansson P, Hedner P, Nilsson-Ehle P (1983) Relations between thyroid function, hepatic and lipoprotein lipase activities and plasma lipoprotein concentrations. Acta Endocrinol (Copenh.) 104:50–56.
22. Wahlquist ML, Fidge NH, Lomas F (1977) Lipoprotein composition in hypothyroidism. Clin Chim Acta 77:269–274.
23. Wegelius O. Connective tissue change. In: Werner SC, Ingbar SH (eds). The Thyroid 3ed 520-527, Harper & Row, New York.
24. Wieland H, Seidel D (1972) Plasmalipoproteine bei Patienten mit Hyperthyreose – Isolierung und Charakterisierung eines abnormen High-Density-Lipoproteins. Z klin Chemie klin. Biochem 10:311–321.
25. Zimmer F, Riebeling V, Benke B, Schuster J, Rosskamm M (1980) Das LDL-HDL-Verhältnis bei Patienten mit Koronarsklerose. Z Kardiol 69:149–153.

Incidence and Treatment of Hyperlipidemia in Primary and Secondary Adrenal Insufficiency

K. EHLENZ, J. STEINSEIFER, H. KAFFARNIK, and A. STEINMETZ

Summary

Although hyperlipidemia occuring in hypopituitarism has been known for a long time, only sporadic reports are available. Recently a survey on 333 patients with hypopituitarism reported a shortened life expectancy in these patients mainly because of vascular disorders. As hyperlipidemia is a major risk factor for vascular disease we investigated the incidence of hyperlipidemia in 23 patients with primary and 35 patients with secondary (due to diseases of the pituitary gland) adrenal insufficiency. All patients had been substituted for deficient hormones according to generally accepted rules. Hypercholesterolemia was the dominant lipid disorder and plasma cholesterol levels above 250 mg/dl occured in more than 60% of the patients studied as compared to 20–30% reported in German and North American populations. 13 patients with severe hyperlipidemia consented to treatment with diet, colestipol or fenofibrate. They responded well and plasma lipid levels could be normalized. The high prevalence of hyperlipidemia may be one major factor for the diminished life expectancy in patients with hypopituitarism (and probably also with primary adrenal insufficiency) and should thus be treated, although the longterm benefit of treatment has not yet been established.

Introduction

Clinical observations on lipid disorders in patients with diseases of the pituitary are very sparce, although as early as 1955 Zöllner [22] observed hyperlipidemia and xanthomatosis in hypopituitarism apparently due to secondary hypothyroidism. Substitution of the patients with thyroid hormone attenuated hypercholesterolemia. Later only few reports apperared dealing with the same phenomenon. In 1961 Jacobs and colleagues [11] reported on late appearence of hyperlipidemia in patients with hypopituitarism. In 1967 Summers et al. (20) published data on lipid disorders in diseases of the pituitary gland. In 1985 a Japanese group focused on the issue again in patients with hypopituitarism [10]. Apparently the most significant lipid abnormality appears in women with Sheehan's syndrome. Accordingly, biochemical studies gave evidence for a lipolytic action of cortisol, ACTH and somatotropine [6]. Also the growth hormone's role in the regulation of hyperlipidemia was studied in detail [3, 7, 14, 21]. Prolonged adrenal cortical stimulation leads to lowering of lipid levels in man [4]. One of the mechanisms responsible for cholesterol lowering by ACTH stimulation in rats was shown to be an induction of LDL receptors in the adrenal gland [13]. The cited publications dealt with only a few patients at a time, no systematic report on subjects with primary and secondary adrenal insufficiency has appeared to our knowledge. Recently in a retrospective study on 333 patients with hypopituitarism an overall higher mortality in this group was found in comparison to a control population [17]. The deaths from vascular disorders were also significantly increased [17]. No effective lipid lowering regimens for the treatment of hyperlipidemia persisting even after sufficient hormonal substitution have been proposed. It was the purpose of the present study to investigate systematically the incidence of hyperlipidemia in primary and secondary adrenal insufficiency and to explore the effectiveness of lipid lowering medication in cases with severe hyperlipidemia.

Patients and Methods

Patients

23 patients with primary adrenal insufficiency and 35 patients with secondary adrenal insufficiency who are seen on a regular basis at the outpatient facility of our department of Endocrinology and Metabolism were investigated. Of the 23 patients with primary adrenal insufficiency 15 suffered from autoimmune and 5 from tuberculotic Addison's disease, 3 had undergone bilateral adrenalectomy for Cushing's disease. Two of the patients with Addison's disease of autoimmune origin were substituted for hypothyroidism (Schmidt-Syndrome). Of the 35 patients with secondary adrenal insufficiency 26 had undergone hypophysectomy for tumor treatment, 4 patients had panhypopituitarism of unknown origin, 5 patients had Sheehan's syndrome. All patients were sufficiently substituted according to generally accepted criteria. There were 31 women and 27 men in the entire group.

Lipid and Lipoprotein Analysis

Blood was drawn after an overnight fast. Serum was obtained after coagulation for 1 hr at room temperature. Serum cholesterol and triglyceride levels were determined enzymatically (Boehringer Mannheim, Germany), HDL cholesterol and LDL cholesterol by quantitative lipoprotein electrophoresis as distributed by Immuno GmbH (Heidelberg, Germany). Apolipoproteins Al and B were quantitated by radial immunodiffusion assay (Immuno GmbH, Heidelberg, Germany).

Statistical Analysis

Parameter between groups were analyzed by the parameterfree test of Wilcoxon, Mann and Whitney or Kruskal-Wallis.

Results

Prevalence of Hyperlipidemia

Lipid and lipoprotein parameters of the two groups of patients with adrenal insufficiency are summarized in table 1. As patients with Sheehan's syndrome exert a very prominent hyperlipidemia they are listed separately as a small

Table 1. Lipid and Lipoprotein parameters in patients with primary and secondary adrenal insufficiency. Sheehan's syndrome as a very prominent lipid affecting disorder is listed as a subgroup of secondary adrenal insufficiency. Values as $\bar{x} \pm SD$

	Group I: primary adrenal insufficiency	Group II: secondary adrenal insufficiency	
		M. Sheehan	other causes
n:	23	5	30
Age (years):	49,0±15,5	53,6± 7,2	55,1±13,6
Cholesterol(md/dl):	241,9±43,6	267,0± 35,8	255,8±56,0
Triglycerides (mg/dl):	161,1±79,2	214,0±100,7	191,0±73,5
HDL-Cholesterol (mg/dl):	57,9±18,6 (n=17)	64,3± 15,1 (n=3)	47,9±12,0 (n=13)
LDL-Cholesterol (mg/dl):	170,0±33,8 (n=16)	230,4± 9,4 (n=3)	206,6±40,4 (n=12)

Table 2. Distribution of patients' serum cholesterol and triglyceride values. Note that hypercholesterolemia is the prominent feature of lipid disorder in hypopituitarism and Addison's disease, leading to type II a and II b hyperlipidemias

		Serum cholesterol levels (mg/dl)			Serum triglyceride levels (mg/dl)			Hyperlipidemia Type	
		>200	200-250	<250	>200	200–250	250–500	II a	II b
Primary adrenal insufficiency									
	n:	4	8	11	15	5	3	11	8
	%	17.4	34.8	47.8	65.2	21.7	13.0	47.8	34.8
Secondary adrenal insufficiency									
– Sheehan's syndrome	n:	–	1	4	2	–	3	2	3
	%	–	20	80	40	–	60	40	60
– other causes	n:	5	7	18	17	8	5	14	11
	%	16.7	23.3	60	56.7	26.7	16.7	46.7	36.7

subgroup. Table 2 shows the attribution of lipid abnormalities to ranges of plasma cholesterol and triglyceride levels. This table does not discriminate between different age groups.

HDL cholesterol levels were not affected in both groups of patients as compared to normal populations, they even tended tu be increased [1, 9]. Mainly hyperlipidemia-phenotypes IIa and IIb according to the Fredrickson classification were seen. We did not establish a direct age and sex-matched control group as this could likely lead to artifacts in a small group. We rather compared the data to results of two big surveys carried out in North America [9] and in Germany [1]. At mean age of 50 to 55 years in the normal population about 20–30% of men and women have serum cholesterol levels above 200 mg/dl. As outlined in tables 1 and 2, they majority of patients with adrenal insufficiency exhibit cholesterol levels above 200 mg/dl. This is clearly demonstrating the high prevalence especially of hypercholesterolemia in the studied population of patients with primary and secondary adrenal insufficiency.

Treatment of Hyperlipidemia

A total of 13 patients with severe hyperlipidemia consented to a drug treatment in addition to dietary regimens. These were 6 patients with primary and 7 patients with secondary adrenal insufficiency. Patients were given colestipol (4 g tid) in type II a hyperlipidemia (total cholesterol above 250 mg/dl, triglycerides below 200 mg/dl) or fenofibrate 200–300 mg q. d. in type II b hyperlipidemia, when triglycerides exceeded 200 mg/dl. Lipoprotein and apoprotein parameters were measured before and after 6 months on treatment. There were no severe side effects that would have led to a discontinuation of the study. No drop-outs were noted, all patients completed the study. In all cases we noted a highly significant lowering of serum cholesterol by 25%, reflecting a drop in LDL-cholesterol by 23% and a triglyceride lowering of 34%. HDL cholesterol did not change significantly. Table 3 summarizes those results .

Table 3. Lipid-, lipoprotein- and apoprotein changes during dietary and drug treatment (colestipol, fenofibrate) in patients with primary (n = 6) and secondary (n = 7) adrenal insufficiency. The table shows data before and after 6 months of treatment (Apo = Apolipoprotein, values as $\bar{x} \pm SD$)

(n = 13)	before therapy (mg/dl)	after therapy (mg/dl)	P
Cholesterol	300.8 ± 41.0	225.9 ± 40.2	0.0015
Triglycerldes	214.4 ± 80.0	143.3 ± 58.8	0.0015
HDL-Cholesterol	60.4 ± 15.6	69.5 ± 15.6	0.7241
LDL-Cholesterol	207.5 ± 41.3	160.7 ± 38.2	0.0022
Apo Al	167.3 ± 25.6	161.4 ± 27.0	0.1235
Apo B	120.3 ± 36.5	90.5 ± 18.9	0.0209

Discussion

A recent study clearly shows that life expectancy in hypopituitarism is shortened because of vascular disorders [17]. There was no evidence to suggest that these patients differed from the normal population with respect to variables such as consumption of alcohol, smoking habits, and ethnic and socioeconomic background [17]. The mortality risk was increased irrespective of the disease the hypopituitarism resulted from; also the degree of pituitary dysfunction did not seem to be important in death hazzard function. Hypertenion and diabetes mellitus also did not seem to influence death hazzard function although the group investigated was relatively small to answer this question [17]. Hypopituitarism, however, is associated with a high prevalence of hyperlipidemia as shown in the present study, confirming more or less sporadic earlier reports [10, 11, 20, 22]. We extend the observation to patients with primary adrenal insufficiency in whom we also found a high prevalence of hyperlipidemia. Both kinds of adrenal insufficiency show mainly an elevation of LDL cholesterol, one of the most atherogenic lipoprotein particles. High LDL cholesterol levels are a major risk factor for the development of atherosclerosis [8]. Rosen and Bengtsson [17] have attributed hyperlipidemia in hypopituitarism to the lack of growth hormone and growth hormone deficiency to explain premature death from vascular disease. It seems apparent that growth hormone deficient patients often have hypercholesterolemia [3, 14] and treatment with growth hormone reduces cholesterol levels [3, 7], restores body composition towards normal and reduces the high waist/hip ratio [18]. However patients with primary adrenal insufficiency do also present a high prevalence of hyperlipidemia despite their apparently intact growth hormone secretion. It is hard to imagine that hyperlipidemias in both kinds of adrenal insufficiency are of an entirely different pathogenesis. Growth hormone deficiency might contribute to hypercholesterolemia and be one of the mechanisms of hyperlipidemia.

We like to extrapolate from our data that primary adrenal insufficiency might also be associated with premature death possibly in part mediated through hyperlipidemia. Other mechanisms that may contribute to hypercholesterolemia include the decay of a high numer of LDL receptors in the adrenal gland, usually expressed by ACTH [13]. Also DHEA could explain some of the hyperlipidemia mortality problems. DHEA is produced in the adrenal gland by ACTH stimulation. DHEA-S has a 20-fold higher serum level than any of the steroid hormones and decreases with age [16]. It was shown earlier that the administration of DHEA leads to a decrease in serum cholesterol, mainly LDL cholesterol, and to a reduction in body fat [12, 15]. In a prospective study in 242 men, aged 50 to 79 years, DHEA-S levels below 140 μg/dl were associated with an increased risk of cardiovascular disease [2]. In summary, hyperlipidemia in primary and secondary adrenal insufficiency seems to be caused by complex mechanisms. The fact that patients with secondary adrenal insufficiency have a somewhat more severe hyperlipidemia than those with primary adrenal insufficiency (data not shown) argues for growth hormone deficiency as an additional cause of hypercholesterolemia.

The increased death rate in patients with hypopituitarism from vascular disorders, in part mediated through hyperlipidemia should justify the treatment of hyperlipidemia in these patients. Treatment of hyperlipidemia in primary and secondary intervention trials in the "normal" population was shown to be associated with a decrease in coronary heart-and peripheral vascular disease [5, 19]. Although no study is available yet to unequivocally proof the benefit of treatment of hyperlipidemia in patients with primary and secondary adrenal disease, we do treat patients who exhibit striking hyperlipidemia. The lipid disorder responds well to colestipol or fenofibrate and treatment is well tolerated at the doses given in this study. The proof, however, of increased mortality risk from vascular disorders caused by hyperlipidemia in primary adrenal insufficiency and the benefit in terms of reduction of this risk by treatment of hyperlipidemia is still lacking and requires further studies.

References

1. Assmann G, Schulte H (1992) Hyperlipidemia according to age. Excerpta Medica Tokio (in press).
2. Barrett-Conner E, Khaw KT, Yen SSC (1986) A prospective study of dehydro epiandrosterone sulfate, mortality, and cardiovascular disease. N Eng J Med 315, 24:1519–1524.
3. Blackett PR, Weech PK, McConathy WJ, Fesmire JD (1982) Growth hormone in the regulation of hyperlipidemia. Metabolism 31:117–120.
4. Conn JW, Vogel WC (1949) Effects of prolonged adrenal cortical stimulation upon free and esterified cholesterol in normal men. J Clin Endocrin 9:656–660.
5. Essen von R (1992) Regression der Koronarsklerose durch Behandlung der Fettstoffwechselstörung? Erste klinische Ergebnisse und methodische Probleme. Internist 33:62–66.
6. Fredrickson DS, Gordon RS Jr (1958) Transport of fatty acids. Phys Rev 38: 585–596.
7. Friedman M, Byers SO, Rosenman RH, Li CH, Neuman R (1974) Effect of sub acute administration of human growth hormone on various serum lipid and hormone levels of hypercholesteremic and normocholesterolemic subjects. Metabolism 23:905–912.
8. Havel RJ (1989) Biology of cholesterol, lipoproteins and atherosclerosis. Clin Exp Hyerth Pract A 11:887–900.
9. Heiss G, Tamir I, Davis CE, Tyroler HA, Rifkand BM, Schonfeld G, Jacobs D, Frantz ID Jr (1980) Lipoprotein-cholesterol distributions in selected North American populations: the lipid research clinics program prevalence study. Circulation 61, 2:302–315.
10. Ishibashi S, Murase T, Yamada N, Tanaka K, Takahu F, Sato K (1985) Hyperlipidaemia in patients with hypopituitarism. Acta Endocrinol 110:456–460.
11. Jacobs RJ, Krieger DT, Charles RN (1961) Late appearance of hyperlipemia in hypopituitarism. Ann Int Med 55:640–646.
12. Kask E (1959) 17-Ketosteroids and Arteriosclerosis. Angiology 10: 358–368.
13. Kovanen PT, Goldstein JL, Chappell DA (1980) Regulation of low density lipoprotein receptors by adrenocorticotropin in the adrenal gland of mice and rats in vivo. J Biol Chem 255:5591–5598.

14. Merimee T, Hollander W, Fineberg E (1972) Studies of hyperlipidemia in the hGH-deficient state. Metabolism 21:1053–1061.
15. Nestler JE, Barlascini CO, Clore JN, Blackard WG (1988) Dehydroepiandrosterone reduces serum low density lipoprotein levels and body fat but does not alter insulin sensitivity in normal men. J Clin Endocrinol Metab 66:57–61.
16. Orentreich N, Brind JL, Rizer RL, Vogelman JH (1984) Age changes and sex differences in serum dehydroepiandrosterone sulfate concentrations throughout adulthood. J Clin Endocrinol Metab 59:551–555.
17. Rosén T, Bengtsson B-Å (1990) Premature mortality due to cardiovascular disease in hypopituitarism. The Lancet 336:285–288.
18. Salomon F, Cuneo RC, Hesp R, Sönksen PH (1989)The effects of treatment with recombinant human growth hormone on body composition and metabolism in adults with growth hormone deficiency. N Engl J Med 321:1797–1803.
19. Steinmetz A, Karrarnik H (1992). Medikamentose Therapie der Hyperlipoproteinamien. Internist 33:44–53.
20. Summers VK, Jipkin LJ, Davis JC (1967) Serum lipids in diseases of the pituitary. Metabolism 16, 12:1106–1113.
21. Thuesen L, Christiansen JS, Sorensen KE, Jorgensen JOL, Orskov H, Henningsen P (1988) Increased myocardial contractility following growth hormone administration in normal man. Dan Med Bull 351:193–196.
22. Zöllner N (1955) Hypopituitarismus (Sheehansches Syndrom) mit Hyperlipämie und Xanthomen. Dt Med Wschr 80:999–1002.

Cholesterol Lowering Effect of Simvastatin in Patients with Normal and Impaired Renal Function

G. Wambach, R. Jacob, E. Danull, and H. Eckardt

Summary

The cholesterol lowering effect of simvastatin was studied in hyperlipidaemic patients with normal and impaired renal function. 10 patients (group A, 4 male/6 female, aged 57±8 years, serum creatinine 0.95±0.10 mg/dl) and 10 patients with chronic interstitial nephritis (group B, 4 male, 6 female, aged 64±6 years, serum creatinine 2.13±0.65 mg/dl) were treated with simvastatin for a period of 12 weeks after a placebo period of 4 weeks. Total cholesterol was reduced from 308±25 to 222±31 mg/dl in group A and from 277±28 to 204±28 mg/dl in group B. Mean final dose of simvastatin was 34 and 31 mg/d respectively. This reduction was seen only in the LDL fraction of cholesterol: in group A from 209±44 mg/dl to 140±30 and in group B from 185±21 to 104±37 mg/dl. HDL cholesterol was unaffected by simvastatin treatment.

In summary, simvastatin effectively lowers total and LDL-cholesterol in patients with normal renal function and in patients with mild chronic renal failure.

Introduction

Cardiovascular diseases are contributing factors for the morbidity and mortality of patients with chronic renal failure. This results primarily from the high incidence of associated cardiovascular risk factors such as hypertension and hypercholesterolemia [2, 5, 8, 10, 11]. Treatment of hypercholesterolemia in patients with chronic renal failure is especially important owing to the high cardiovascular risk involved. However, therapeutic possibilities are limited by the impaired renal elimination of many lipid lowering drugs. The HMG-Co-A-reductase inhibitor simvastatin, however, is eliminated mainly via the liver. This opens up new avenues of treatment for patients with chronic renal failure and secondary hyperlipidemia. The present study compared the efficacy and tolerance of simvastatin in patients with normal and impaired renal function.

Patients and Protocol

The study included patients with a total cholesterol level of over 250 mg/dl and patients with a total cholesterol level of over 200 mg/dl receiving lipid lowering medication.

Group A included ten normorenal patients, 4 men and 6 women with a mean age of 57±8 years. Four patients (one woman, three man) had already suffered from a myocardial infarction, while the other six had no signs of coronary heart disease. The renal functional parameters were within the normal range in all the patients, the serum creatinine level being 0.96±0.1 mg/dl and the endogenous creatinine clearance 103.6±12.4 ml/min.

Group B, consisted of 10 patients with renal failure due to interstitial nephritis, 4 men and 6 women with a mean age of 64±6 years with a serum creatinine level of 2.13±0.65 mg/dl. The endogenous creatinine clearance was limited to 41.8±7.7 ml/min. Three patients (one woman, two men) had already suffered from a myocardial infarction; three other patients suffered from angina pectoris.

The study was divided into three main arms, a screening phase, a placebo phase and a three-month therapeutic phase. In accordance with the guidelines of the American Heart Association the objective of the change in diet was a maximum fat intake accounting for 35% of the daily caloric requirements. The proportion of unsaturated fatty acids exceeded that of saturated fatty acids, and the daily intake of cholesterol was limited to 300 mg.

The screening phase was followed by a four-week placebo phase, during which no lipid lowering drugs were administered.

In the final treatment period the initial dose of 10 mg simvastatin was increased to 20 mg and then to 40 mg at four-week intervals if the total serum cholesterol level still exceeded 200 mg/dl.

Results

At the end of the placebo period, total cholesterol averaged 308.3±24.8 mg/dl in patients with normal renal function (A) and 277.2±28.5 mg/dl in renal failure group (B). In group A the cholesterol fell to 243±34 mg/dl and group B to 208±38 mg/dl at the end of the 4 weeks.

In both groups LDL-cholesterol levels fell most markedly in parallel with the total cholesterol level during the first treatment phase. In the group of normorenal patients (A) LDL-cholesterol fell from 209±44 to 165±38 mg/dl, in group B from 185±22 to 129±24 mg/dl.

HDL-cholesterol remained constant in both groups. During the first 4 weeks, HDL-cholesterol in group A was 39±13 and 40±14 mg/dl and in group B 48±23 and 49±27 mg/dl.

During the course of the study a further reduction in total and LDL-cholesterol was achieved with simvastatin doses of 20 mg beginning on day 28 and 40 mg beginning on day 56 in those patients with serum cholesterol val-

Table 1. Total cholesterol and LDL-cholesterol ($\bar{x} \pm$ SEM)

		Day 1	Day 28	Day 56	Day 84
Group A	Total cholesterol (mg/dl)	308±24	243±34	230±26	222±31
	LDL (mg/dl)	209±44	165±39	152±31	140±30
Group B	Total cholesterol (mg/dl)	277±28	218±32	220±20	204±28
	LDL (mg/dl)	185±21	129±24	120±30	104±37

ues still above 200 mg/dl (Table 1). Mean daily dose of simvastatin was 34±10 mg in group A and 31±1 mg in group B. At the end of the study total cholesterol was lowered to 222±31 mg/dl in group A and to 204±28 mg/dl in group B. LDL-cholesterol was reduced to 140±30 mg/dl in group A and to 105±39 mg/dl in group B. HDL-cholesterol remained constant in both groups.

Over the entire course of the study the triglyceride values were not changed in both groups treated with simvastatin.

Discussion

The majority of large epidemiologic studies in Europe and the United States have shown that the risk of coronary heart grows linearly with the total cholesterol level in the blood [5, 6, 11, 12]. The decisive factor for the formation and development of arteriosclerotic foci in arterial vessels is the proportion of circulating LDL-cholesterol in the blood. The more LDL there is in the blood, the more rapidly arteriosclerosis develops [4].

Cholesterol metabolism in individual cells is regulated by LDL receptors on the cell surface. These take up LDL-cholesterol from the circulating blood, break down the lipoprotein and utilize the released cholesterol. 70% of the entire LDL is metabolized by the liver.

The number of LDL-receptors of the single cell is regulated by the concentration of cholesterol inside of the cells. When the cell is saturated with cholesterol, the number of receptors on the cell surface decreases and only a small amount of cholesterol is taken out of the circulation. The opposite applies, the LDL-receptor concentration goes up, when little cholesterol is circulating.

The genesis of Familial Hypercholesterolemia has been largely clarified. A genetic defect associated with a deficiency of LDL receptor activity on the cells may lead at an early point to high blood LDL levels of between 200 and 400 mg/dl in heterotygous patients [13]. This form of Familial Hypercholesterolemia occurs in about 0.2% of the people in western industrialized countries. The rare homozygotic form of hypercholesterolemia with a total receptor deficiency results in LDL blood levels of over 600 mg/dl and if untreated may lead to myocardial infarction and death even in childhood.

In contrast, the genesis of secondary hypercholesterolemia due to chronic renal failure is for the most part still unclear. Reduced activity of lipoprotein lipase, hyperinsulinemia, parathormone hypersecretion and carnitine deficiency have all been advanced as possible factors.

Lipid lowering substances like the fibrates are eliminated mainly by the kidneys. For example, 70% of gemfibrozil is eliminated renally; clofibrate has a halflife of 17.2 hours in patients with normal renal function, whereas patients undergoing dialysis show halflife of between 90 and 180 hours [4]. Bezafibrate with a halflife of 2.1 hours showed an average 4 times higher halflife of 8.5 hours in patients with impaired renal function [3]. Therefore, fibrates must be administered in reduced doses to patients with impaired renal function.

The metabolism of simvastatin and lovastatin is independent of the kidneys [1]. About 83% of the substance is excreted with the bile, and only 10% is eliminated by the kidneys.

The present study confirmed the good tolerance profile and efficacy of simvastatin in patients with hyperlipidemia and normal renal function. The results in terms of efficacy and tolerance did not differ between patients with mild to moderate renal failure and patients with normal renal function. A degree of renal failure marked by a 60% reduction in glomerular filtration rate does not necessitate dosage reduction.

Simvastatin therefore permits effective reduction of cholesterol levels with few adverse reactions in patients with hypercholesterolemia secondary to renal failure. However, long-term prospective studies must clarify whether this therapy can also lead to a reduction in the cardiovascular morbidity and mortality in patients with chronic renal failure.

References

1. Alberts AW, Chen J, Kuron G (1980) Mevinolin: a highly potent competitive inhibitor of 3-hydroxy-3-methylglutaryl coenzyme A reductase, and a cholesterol lowering agent. Proc Natl Acad Sci USA 77:3957–3961.
2. American Heart Association Steering Committee Medical and Community Program (1980) Risk factors and coronary disease. A statement for physicians. Circulation 62:449.
3. Anderson P, Norbeck HE (1981) Clinical pharmacokinetics of bezafibrate in patients with impaired renal function. Eur J Clin Pharmac 21:209.
4. Brown MS, Goldstein JL (1985) Arteriosklerose und Cholesterin: die Rolle der LDL Rezeptoren. Spektrum der Wissenschaft 1:96–106.
5. Dawber TR (1980) The Framingham Study. The Epidemiology of Atherosclerotic Disease (Harvard University Press: Cambridge).
6. Faed EM, McQueen EG (1979) Plasma half life of clofibric acid in renal failure. Br J Clin Pharmac 7:407.
7. Helber A, Wambach G, Bottcher W, Weller P, Schmidt R (1980) Hypercholesterolaemia and hypertriglyceridaemia in patients with analgesic nephropathy. Nephron 26:111–115.

8. Helber A, Wambach G, Böttcher W, Finke K, Hahn R (1981) Erhöhtes kardiovaskulares Arterioskleroserisiko bei Analgetika-Nephropathie. Dtsch Med Wschr 106:1369–1373.
9. Lipid Research Clinic Program (1984) The Lipid Research Clinic Coronary Primary Prevention Trial Results. II. The Relationship of reduction in incidence of coronary heart disease to cholesterol lowering. J Amer Med Ass 251:365.
10. Manninen V, Elo MO, Frick MH, Haapa K, Heinonen OP, Heinsalmi P, Helo P Hutunnen JK, Kaitaniemi P, Koskinen P, Mäenpää H, Mälkönen M, Mänttäri M, Norola S, Pasternack A, Pikkarainen J, Romo M, Sjöblom T, Nikkilä EA (1988) Lipid alterations and decline in the incidence of coronary heart disease in the Helsinki Heart study. J Amer med Ass 260:641.
11. Steinberg D and the NIH Consensus Conference (1985) Lowering blood cholesterol to prevent heart disease. J Amer med Ass 253:2080.
12. Strategies for the prevention of coronary heart disease a policy statement of the European Atherosclerosis Society. Eur Heart J 8:77–88 (1987).
13. Tyroler HA (1987) Lowering plasma cholesterol levels decrease risk of coronary heart disease: an overview of clinical trials. Hypercholesterolemia and atherosclerosis: pathogenesis and prevention. In: Steinberg D, Olefsky JM, eds. Contemporary Issues in Endocrinology and Metabolism. Vol 3 New York: Churchill Livingstone, 99–115.

Determinants for Coronary Atherosclerosis Regression. The Marburg Lipid Intervention Trial

M. STELLWAAG, B. MAISCH, B. E. STRAUER, M. LESCHKE, J. SCHNEIDER, A. STEINMETZ, and H. KAFFARNIK

Summary

In order to find out whether diet and drug therapy of hyperlipidemia changes the course and outcome of coronary artery desease, patients with angiographically proven coronary artery disease and hyperlipidemia were treated with diet, fenofibrate and colestipol. By reangiography after two to four years three groups of patients could be formed. Those which showed a progression, no change or a regression of coronary artery disease (CAD). These groups had different levels of lipoproteins under therapy. The regression group had lowest and the no change group had lower levels of total cholesterol and low density lipoprotein (LDL)-cholesterol as compared to the progression group. High density lipoprotein (HDL)- cholesterol had an inverse correlation. These results suggest that a lipid lowering therapy improves course and outcome of CAD.

Introduction

First evidence for the possibility of atherosclerosis regression came from observations by Anitschkow [1] in the early thirtees who noted that it was necessary to continously feed cholesterol to rabbits in order to maintain experimentally induced lesions. Meanwhile a whole body of evidence from experimental animal studies has proven the possibility of atherosclerosis regression [2, 10, 8, 14]. Human studies designed at investigating atherosclerosis regression initially were based on protocols including lipid lowering agents that produced small changes in blood lipid levels [15, 9]. In 1987 the Cholesterol Lowering Atherosclerosis Study (CLAS) [3] introduced aggressive lipid lowering by niacin and colestipol and resulted in a significant coronary atherosclerosis regression and a slow down of progression. Since then a series of investigations aimed at retarding progression of CAD have been published. The Lifestyle Heart Trial [16] intervened by asking participants to make changes in lifestyle (increase exercise time, decrease dietary fat and cholesterol, reduce calories). The experimental group of patients experienced a reduction in average percentage diameter of coronary stenosis from 40.0 to 37.8 as compared to controls who showed an increase from 42.7 to 46.1. Also in 1990 the results of the POSCH study (Program on Surgical Control of the Hyperlipidemias) were reported [6]. Lipid lowering was achieved by partial ileal bypass surgery on patients that had survived a first myocardial infarction. It was found that partial ileal bypass reduces the morbidity from CAD in patients who have had a myocardial infarction. Recently several other secondary intervention trials in coronary artery atherosclerosis not only in males but also in females [13] were reported using different lipid lowering drugs. These drugs included niacin-colestipol, niacin-lovastatin [5, 13], niacin and colestipol [3, 7] and fenofibrate [11]. In these studies an unequivocal benefit of lipid lowering therapy in terms of slowing down progression or even promoting regression of CAD could be documented.

The purpose of this study was to find out whether treatment of hyperlipoproteinemia changes the course and outcome of a coronary artery disease in humans. In addition it should be examined which lipoproteins are determinants for the progression of CAD.

A further question of our study was to examine the contribution of testosterone to the course of CAD under lipid lowering therapy. This question was based on results of Braunsteiner et al. [4]. They found a very strong inverse association between coronary heart desease and lipoprotein lipase (LPL)-activity on one hand and a positive correlation between LPL-activity and plasma testosterone on the other hand in order to identify a primary alteration of lipid metabolism which could be responsible for the development of coronary heart desease. With these results in mind we measured testosterone in males to relate male sex hormone to the course of CAD under lipid lowering therapy.

Patients and Methods

The study was started 1986, when 87 patients with angiographically proven CAD and hyperlipoproteinemia type II a, n=19, type II b, n=14, and type IV, n=3 were initially enrolled. Meanwhile 36 patients (30 males and 6 females, age between 41 and 72, mean 59 years, S_x=8.8) completed an observation period of 3–4 years and underwent repeat coronary angiography. Their data are reported.

During the observation period the patients had been treated by weight reduction, diet, fenofibrate 250 mg per day and/or colestipol up to 4 packets per day. Before and during therapy four times a year total cholesterol, LDL cholesterol, HDL cholesterol, LDL/HDL ratio and triglycerides were measured. In males testosterone was measured before the initiation of therapy. The coronary angiograms were analysed by two experienced cardiologists after blinding the angiograms to their sequence and to the patients' identity. By measurement of the percent diameter of the stenosis before and after therapy three groups of patients could be formed: progression of CAD, no change of CAD and regression of CAD. In these groups the changes of lipoprotein levels before and during therapy and testosterone levels were compared. After testing for the homogenity of variances (Fisher test) statistical analysis was performed by Students t-test and Wilcoxon U-test.

Results

By comparison of the vessels' narrowing before and during treatment of hyperlipidemia patients were grouped in three categories. In 12 patients progression of CAD was observed, in 21 patients no change of CAD was present and in even 3 patients there was a regression of CAD. Coronary artery segments grafted by a coronary bypass were not considered. The three groups of patients with different courses of CAD differ in their lipoprotein levels as discribed below.

Total Cholesterol

During treatment of hyperlipidemia the serum levels of total cholesterol decreased in the three groups. In the no change group and in the regression group this decrease is statistically significant. But there is no significant difference between total cholesterol levels of the three groups during treatment. The data are demonstrated in Fig. 1.

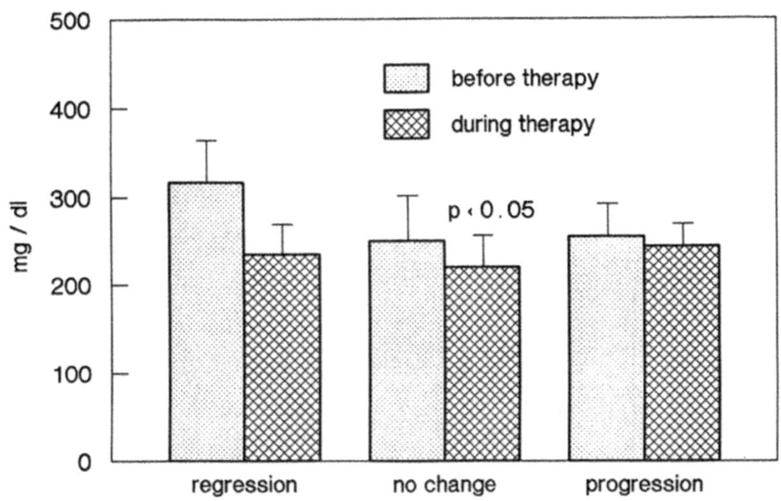

Fig. 1. Total serum cholesterol levels in patients before and during lipid lowering therapy. Three groups were formed according to angiographical evaluation of coronary vessels before and during therapy (regression of CAD n = 3, no change n = 21, progression n = 12)

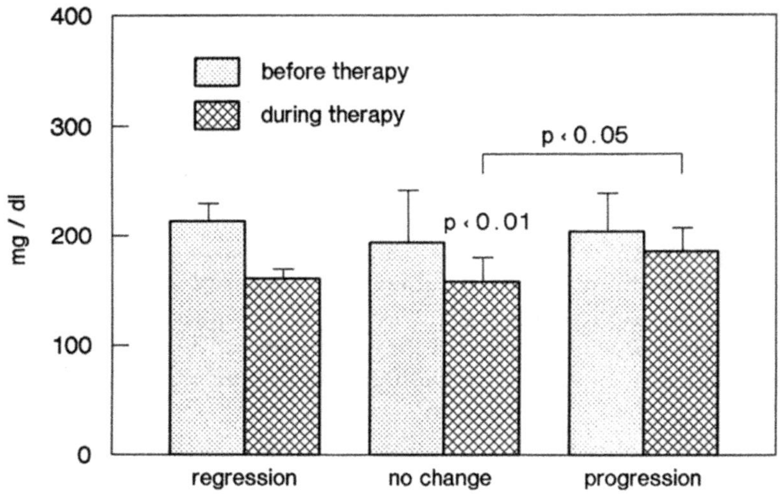

Fig. 2. LDL cholesterol levels before and during therapy. Three groups and formed as explained in Fig. 1

LDL-Cholesterol

Pretreatment LDL-cholesterol levels did not differ between the groups. Only in the "no change" group LDL-cholesterol decreased significantly and differs from LDL-cholesterol in the progression group after therapy (Fig. 2).

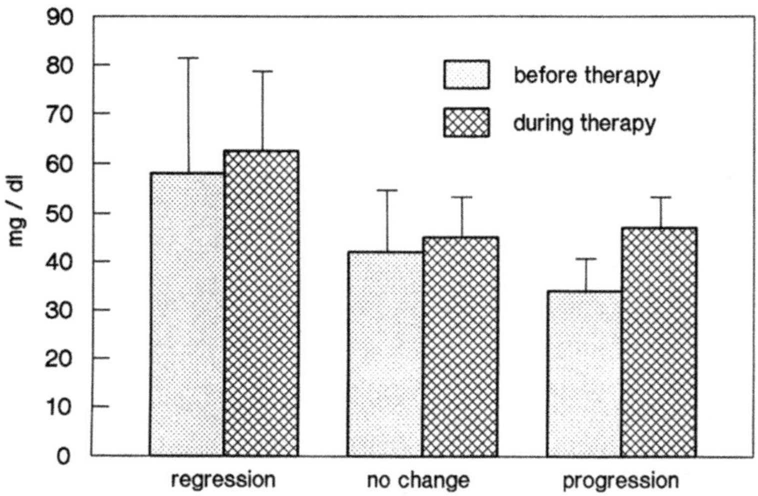

Fig. 3. HDL cholesterol levels before and during therapy. Three groups are formed as explained in Fig. 1

HDL-Cholesterol

During treatment HDL-cholesterol levels increased, but the difference before and during treatment were not statistically significant (Fig. 3).

LDL/HDL Ratio

The most obvious difference in the three groups was seen in the LDL/HDL ratio. The ratio decreased in the three groups during treatment of hyperlipidemia. In the "no change" group this decrease was significant. Furthermore the LDL/HDL ratio during treatment was significantly lower in the "no change" group than in the "progression" group (Fig. 4).

Triglycerides

Triglyceride levels were lowered during treatment of hyperlipidemia. But the changes in the three groups were not statistically significant (Fig. 5).

Testosterone

Patients in the "no change" group showed a statistically significant higher serum level of testosterone than in the progression group (Fig. 6).

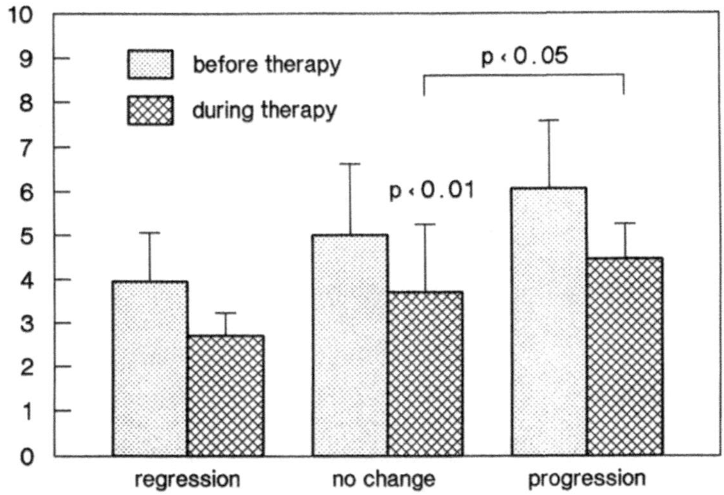

Fig. 4. LDL/HDL Ratio before and during therapy. Three groups are formed as explained in Fig. 1

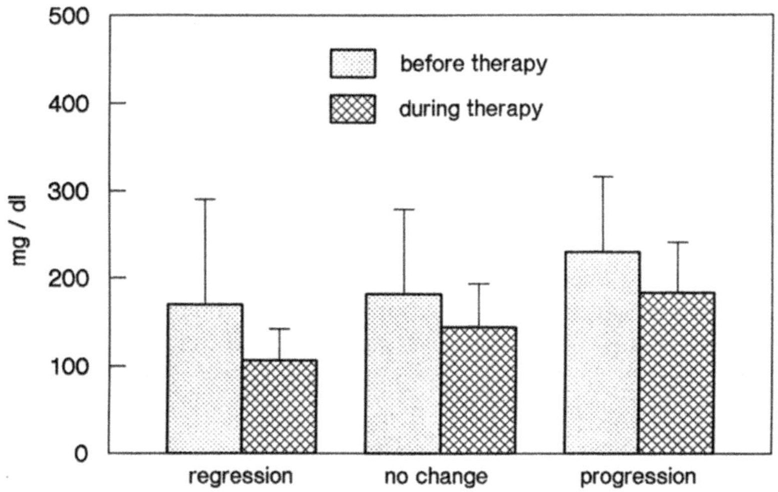

Fig. 5. Triglycerides before and during therapy. Three groups are formed as explained in Fig. 1

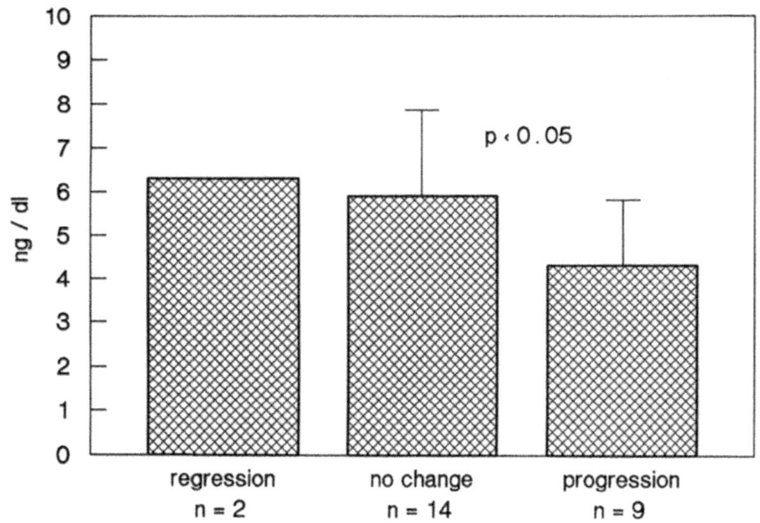

Fig. 6. Testosterone levels before therapy. Three groups are formed as explained in Fig. 1

Discussion

The progression of CAD is slowed down by effective treatment of hyperlipidemia. In some cases even a regression of CAD can be observed. These results are consistent with the results of Blankenhorn et al. [3], Brown et al. [5], Cushin-Hemphill et al. [7], Hahmann et al. [11], and Kane et al. [13]. Our study differs from these studies in that these investigators introduced a placebo or a mildly treated group. While in our study because of ethical reasons hyperlipidemia in all patients was treated by the same effective regime. We assumed that patients not adhering to the treatment would form an "internal control group". The final differences in the effect of treatment i.e. the lower response to this treament in one part of the patients would be either due to a lower compliance of these patients or to a more therapy resistent hyperlipidemia. It is important that more than half of the patients initially enrolled are not yet reangiographed. These patients did not agree to reangiography in part because of the risk and, more interesting, of experiencing no chest pain. We assume that an enrollement of these patients still to undergo reangiography might increase the percentage of the "no change" subjects and perhaps the "regression" group also.

According to our results lowering of total cholesterol and LDL-cholesterol and especially a decrease of the LDL/HDL-ratio are determinants for a favourable course of CAD. Thus we assume that the aim of a lipid lowering therapy in patients with CAD and hyperlipidemia should be a lowering of cholesterol especially of LDL cholesterol to levels lower than 120–110 mg/dl [17]. Although we could not achieve this aim in our study, the results demonstrate that patients with lowest LDL cholesterol had gained most benefit from therapy.

A higher testosterone level seems to be associated with a better outcome of CAD during treatment of hyperlipidemia.These results are consistent with those of Braunsteiner et al. [4], who found an inverse correlation between CAD and the LPL-activity and furthermore a positive correlation between testosterone and the LPL-activity. Epidemiologic studies [12] however found both: higher testosterone levels in patients with myocardial infarction and lower levels in patients with CAD. The mechanism of this effect is not yet clear. All of our patients even those with high testosterone-levels had developed CAD. But during treatment of hyperlipidemia patients with higher testosterone-levels had an apparently higher benefit from lipid lowering therapy in terms of their CAD. Perhaps testosterone induces a better response to lipid lowering therapy. To answer this question further studies are nessecary.

Acknowledgement

We thank Holphar-Fournier, Sulzbach, Germany for kindly supporting the present study. We acknowledge the skilful technical assistance of Ulrike Otte, Luzie Trier, Achim Gorski, and Ingeborg Muttschall.

References

1. Anitschkow N (1933) A history of experimentation on arterial atherosclerosis in animals: In Atherosclerosis: A Survey of the Problem. Edited by Cowdry EV. New York: McMillan, 271–322.

2. Armstrong ML, Warner ED, Conner WE (1970) Regression of coronary atheromatosis in rhesus monkeys. Circ Res 27:59.
3. Blankenhorn DH, Nessim SA, Johnson RL, Sanmarco ME, Azen SP, Cashin-Hemphill L (1987) Beneficial effects of combined colestipol-niacin therapy on coronary atherosclerosis and coronary venous bypass grafts. JAMA 275: 3233–3240.
4. Braunsteiner H, Breier CH, Mühlberger V, Drexel H, Herold M, Lisch JH, Knapp E (1985) Essential role of post-heparin lipoprotein lipase activity and of plasma testosterone in coronary artery disease. Lancet I:1242–1243.
5. Brown G, Albers JJ, Fisher LD, Schaefer SM, Lin JT, Kaplan C, Zhao XQ, Bisson BD, Fitzpatrick VF, Dodge HT (1990) Regression of coronary artery disease as a result of intensive lipid-lowering therapy in men with high levels of apolipoprotein B. N Engl J Med 323:1289–1298.
6. Buchwald H, Varco RL, Matts JP, Long JM, Fitch LL, Campbell GS, Pearce MB, Yellin AE, Emiston WA, Smink RD, Sawin HS Jr, Campos CT, Hansen BJ, Aip Tuna RN, Karnegis J, Sanmarco ME, Amplatz K, Castaneda-Zuniga WR, Hunter DW, Bissett JK, Weber FJ, Stevenson JW, Leon, AS, Chalmers TC, Posch Group (1990) Effect of partial ileal bypass surgery on mortality and morbitity from coronary heart disease in patients with hypercholesterolemia, report of the program on the surgical control of the hyperlipidemias (POSCH). N Engl J Med 323:946–955.
7. Cashin-Hemphill L, Ack WJ, Pogoda JM, Sanmarco ME, Azen SP, Blankenhorn DH (1990) Beneficial effects of colestipol-niacin on coronary atherosclerosis. A 4-year follow up. JAMA 264:3013–3017.
8. Clarkson TB, Bond MG, Bullock BC, McLaughlin KJ, Sawjer JK (1984) A study of atherosclerosis regression in macaca mulatta. V. Changes in abdominal aorta and carotid and coronary arteries from animals with atherosclerosis induced for 38 months and then regressed for 24 or 48 months at plasma cholesterol concentrations of 300 mg/dl. Exp Mol Pathol 41:96–118.
9. Cohn K, Sakai FJ, Langston MF Jr (1975) Effect of clofibrate on progression of coronary desease: a prospective angiographic study in man. Am Heart J 89: 591–589.
10. Faggiotto A, Ross R, Harker L (1984) Studies of hypercholesterolemia in the non-human primate. I. Changes that lead to fatty streak formations. Arteriosclerosis 4:323–328.
11. Hahmann HW, Bunte TH, Hellwig N, Hau U, Becker D, Dyckmans J, Keller HE, Schieffer HJ (1991) Progression and regression of minor coronary arterial narrowings by quantitative angiography after fenofibrate therapy. Am J Cardiol 67: 957–961.
12. Kalin MF, Zumhoff B (1990) Sex hormones and coronary disease: a review of the clinical studies. Steroids 55:330–352.
13. Kane JP, Malloy MJ, Ports TA, Phillips Nr, Diehl JC, Havel RJ (1990) Regression of coronary atherosclerosis during treatment of familial hypercholesterolemia with combined drug regimens. JAMA 264:3007–3012.
14. Kramsch DM, Aspen AJ, Abramowitz BM, Kreimendahl T, Hood WR Jr (1981) Reduction of coronary atherosclerosis by moderate conditioning exercise in monkeys on an atherogenic diet. N Engl J Med 305:1483–1489.
15. Nash DT, Gensini G, Esente P (1982) Effect of lipid lowering therapy on the progression of coronary atherosclerosis assessed by scheduled repetitive coronary arteriography. Int J Cardiol 2:43–45.
16. Ornish D, Brown SE, Scherwitz LW, Billings JH, Armstrong WT, Ports TA, McLanahan M, Kirkeeide RL, Brand RJ, Gould KI (1990) Can lifestyle changes reverse coronary heart desease? The lifestyle heart trial. Lancet 336:129–133.
17. Steinmetz A, Kaffarnik H (1992) Medikamentöse Therapie der Hyperlipoproteinaemien. Internist 33:44–53.

Medical Imaging System and Their Use in the Diagnosis of Cerebral Attacks Caused by Metabolic Disorders

L. WEBER, S. BRAITINGER, J. ZEHNER, and E. PLACH

Summary

We were looking for vascular alterations of the brain due to risk factors like hypertension or hyperlipidemia. The value of clinical and laboratory findings was investigated as well as the use of several medical imaging systems. According to its high sensitivity in detecting microvascular changes, MRI ist superior to CT and scintigraphy. Vascular alteration in the area of medulla oblongata show good coincidence with risk factors and metabolic disorders.

Introduction

Beside clinical investigation, doppler sonography and digital subtraction angiography, MRI-studies allow us like scintigraphy, to get further informations about the cerebral blood flow and a close approach to cerebral disorder. Using this new imaging system we frequently found changes in the area of medulla oblongata. These findings are called vascular leucencephalopathie or "gliöse Narben". To get more information about these nonspecific findings we startes the following study. We wanted to know, whether there is a coincidence between risk factors and vascular alterations of the brain.

Design of the Study

In a retrospecive study (01.07.85–31.01.87) we investigated 100 patients with cerebral vascular lesions. In all patients clinical course and laboratory findings were evaluated. We had a look at the risk factors [2, 3, 13, 15] and compared the different medical imaging systems. In 40 patients simultaneously CT, pulsed doppler sonography, scintigraphy of the brain and digital subtraction angiography was performed. We did a computarizised analysis of the data (Fig. 1).

Results

33% of the patients with vascular alterations were hypertensive, 22% were smokers, 26% suffered from hyperlipidemia, 22% from diabetes and 33%

```
**** Datensatz 86 aus DATEN *****

MR Nr.  2,220 Name SO Geb.Dat. 02/05/25 Alter 61 Geschlecht mannlich

Laborwerte:
Erys    5.25     Hb     16.20     Hkt        48 %
Quick   100      PTT    33        Thrombos   172 000
Crea    0.98     Na     148       K          4.50
HS      6        HST    45        Cholest    200
Triglyc 59       yGt    16        D7         105

Anamnese 1:              Schwache der Extremitaten
Anamnese 2:              Sprachstorungen
Anamnese 3:              Sensibilitatsstorung
Risikofaktoren:
Einweisungsdiagnose:     Apoplex
Entlassungsdiagnose:     Apoplex
Befunde:
MR:   Inf.re.parietal mit geringem Umgebungsodem
CT:   keine Blutung,ausgedehnte Inf-Zone re.parietal bis frontal
DOP:  kein Stromungshindernis der extrakraniellen GefaSe
SZ1:
NEU:
```

Fig. 1. The image showes the data which were recorded in the computerassisted analysis

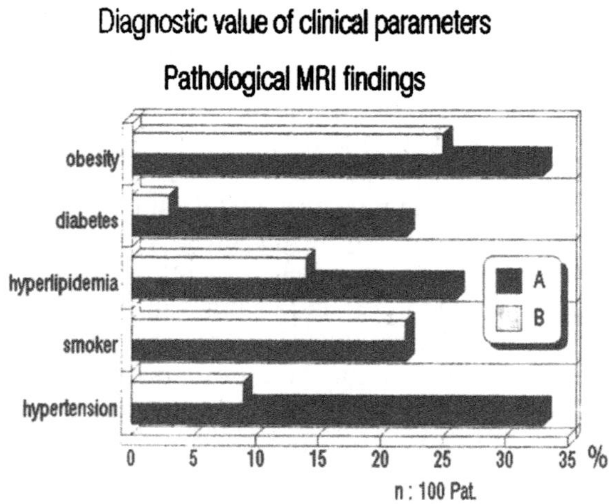

Fig. 2. The graph showes the coincidence between pathological MRI findings and risk factors. The stuffed columns (A) show the percentage of pathological MRI findings in patients with risk factors. The shaded collumns (B) belong to healthy persons

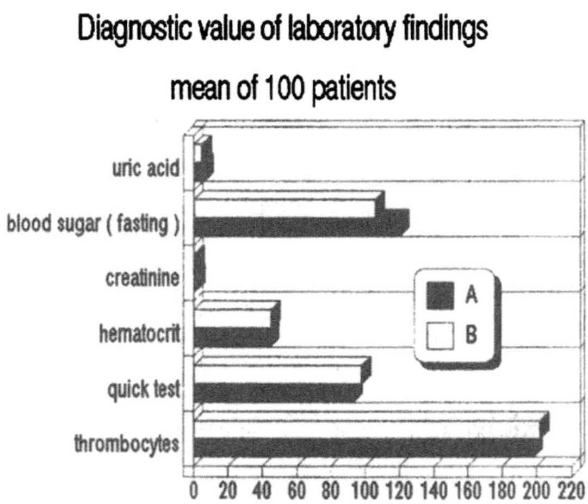

Fig. 3. The graph showes that laboratory findings do not have a digagnostic value in the prediction of cerebral alterations. The stuffed columns (A) bellong to patients with vascular leucencephalopathie. The shaded collumns (B) are due to patients without vascular alterations

were obese. On the other hand 30 patients with normal MRI state were only in 9% hypertensive and in 22% smokers. These persons had in 14% a hyperlipidemia and in 3% diabetes mellitus. 25% were overweight (Fig. 2). There was no coincidence between the laboratory findings and the vascular changes found in the area of medulla oblongata except for lipids and blood sugar (Fig. 3). There are differences in the value of the different imaging systems [1, 4, 5]. Looking for vascular alterations of the brain there is only good coincidence between CT, scintigraphy and MRI (Fig. 4).

Discussions

In patients with risk factors [2, 11, 12, 15] we frequently found vascular alterations in the white matter area and in the area of medulla oblongata. These MRI findings are called vascular leucencephalopathie. Although findings like these are also caused by autoimmune diseases or viral infections there seems to be a close coincidence with metabolic discorders. We suppose that these

Coincidence

▪ MRI / CT	:	62,5 %
▪ MRI / doppler sonography	:	38.5 %
▪ MRI / szintigraphy	:	62,5 %

Fig. 4. There is good coincidence between MRI, CT and scintigraphy in the diagnosis of vascular alterations

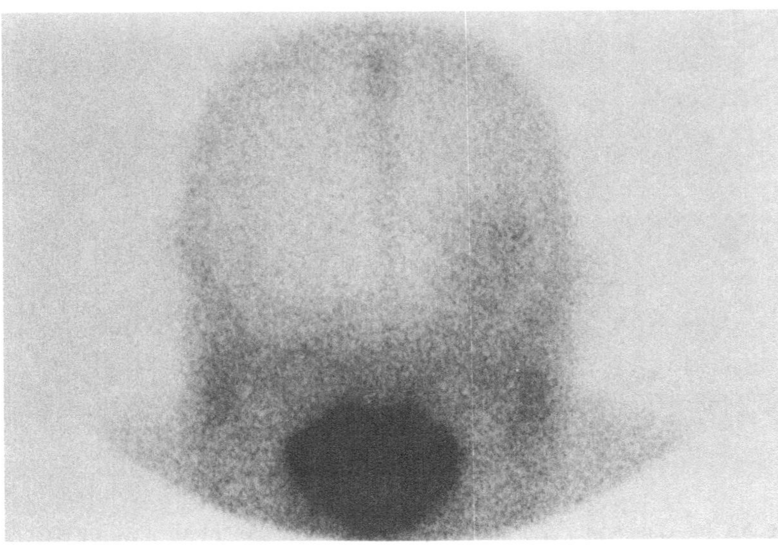

Fig. 5. Scintigraphy of vascular alteration of the left hemisphere

findings are a nonspecific marker of dysfunction in the peripheral vascular systems of the brain. Sonography shows the situation in the extracranial vessels, but it is not able to estimate the cerebral blood flow [14]. Scintigraphy can estimate the cerebral blood flow [8, 10, 16], but cannot differentiate between, ischemia hemorrhage or metabolic reasons of cerebral attacks (Fig. 5). MRI is superior to CT in the diagnosis of microinsults, especially in the region of the brain stem [6, 7, 17, 18, 19] and in case of early detection (Fig. 6 a, b).

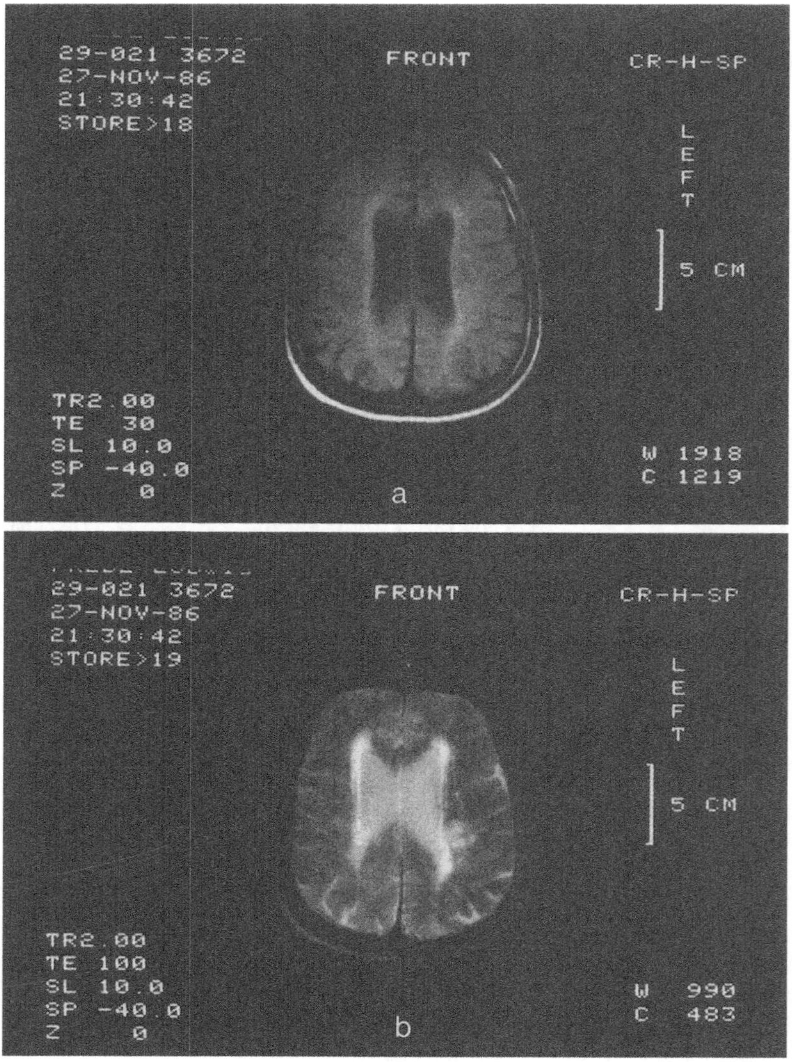

Fig. 6 a, b. Axial spin echo images, proton density weighted, showing periventricular signal hyperintensity (**a**) and acute ischaemia alteration of the left hemisphere next day (**b**)

Conclusion

According to its high sensivity in detecting microvascular changes MRI is superior to other medical imaging systems [4, 5]. Vascular alterations in the area of medulla oblongata called "gliöse Narben", show good coincidence with risk factors like hypertension or hyperlipidemia, but further investigations are necessary [9].

References

1. Becker H, Schwarzrock R, Friedrich H, Hundeshagen H (1977) MR- u. CT-Untersuchungen bei Hirninfarkten im Akut- u. Spätstadium. Thieme, Stuttgart New York.
2. Berkson DM, Stamler J (1965) Epidemiological findings on cerebrovascular diseases and there implications. J Atheroscler Res 5:189–202.
3. Bock KD (1975) Die Differentialdiagnose des Hochdrucks. Ein Basisprogramm für die Praxis. Med Klin 70:67.
4. Bradley WG, Waluch V, Qadley R, Wycoff RR (1984) Comparison of CT and MR in 400 patients with suspected disease of the brain and cervical spinal cord. J Radiology 152:695–702.
5. Brant-Zawadski M, Davis PL, Crooks LE, Mills CM, Norman D, Newton TH, Seldon P, Kaufman L (1983) NMR demonstation of cerebral abnomalities: Comparison with CT. AJR 140:847–854.
6. Bryan RN, Willcott MR, Schneiders NJ, Ford JJ, Derman HS (1983) Nuclear Magnetic Resonance evaluation of stroke. Radiology 149:189–192.
7. Bydder GM, Steiner RE, Young IR, Hall AS, Thomas DJ, Marshall J, Pallis CA, Legg NJ (1982) Clinical NMR imging of brain: 140 Cases. AJR 139:215–236
8. Emrich D (1976) Nuklearmedizinische Diagnostik u. Therapie. Thieme, Stuttgart.
9. Gifford RW Jr (1969) Evaluation of the hypertensive patient with emphasis on detecting curabel causes. Milbank Mem Fund 47:170–186.
10. Hermann HJ (1982) Nuklarmedizin. Urban & Schwarzenberg, München Wien Baltimore.
11. Holme I, Ener SC, Helgeland A, Hjerman I, Leren P, Lund-Larsen PG, Solberg LA, Strong JP (1981) Risk factors and raised atherosclerotic lesion of coronary and cerebral arteries, statistical analysis from the Oslo-Study. Arteriosclerosis 1:250–256.
12. Inter-Society Commission for Heart Disease Resources (1970) Arterosclerosis Study Group: Primary prevention of atherosclerotic diseases. Circulation 42: A55–A95.
13. Julius S, Hansson LH (1983) Borderline hypertension: epidemiologic and clinical implication. In: Genest J, Kuchel O, Hamet P, Cantin M (eds) Hypertension. McGraw-Hill, New York, p 753.
14. Mann WJ (1984) Ultraschall im Kopf-Hals-Bereich. Springer, Berlin Heidelberg New York Tokyo.
15. Marmot MG, Syme SL, Kagan A, Kato H, Cohen JB, Belsky J (1975) Epidemiologic studies of coronary heart disease and stroke in Japanese men living in Japan, Hawaii and California: Prevalence of coronary and hypertensive heart disease and associated risk factors. Am J Epidemiol. 102:514–525.
16. Piaszek L (1975) Die Hirnszintigraphie; Nuklearmedizinische Diagnostik. Klinikarzt 4, 6:235–242.
17. Rink PA, Petersen SB, Muller RN (1986) Magnetresonanz-Imaging u. -Spektroskopie in der Medizin. Thieme, Stuttgart New York.

18. Sipponen JT, Kaste Markku, Ketonen Leena, Sepponen RE, Katevuo Kalevi, Sivula Arto (1983) Serial Nuclear Magnetic Resonance (NMR) imaging in patient with cerebral infarction. J Comp Assist Tomograph 7:585–589.
19. Wende S, Thelen M (1983) Kernspin-Tomographie in der Medizin. Springer, Berlin Heidelberg New York Tokyo.

GPSR Compliance

The European Union's (EU) General Product Safety Regulation (GPSR) is a set of rules that requires consumer products to be safe and our obligations to ensure this.

If you have any concerns about our products, you can contact us on ProductSafety@springernature.com

In case Publisher is established outside the EU, the EU authorized representative is:

Springer Nature Customer Service Center GmbH
Europaplatz 3
69115 Heidelberg, Germany

Batch number: 10583998

Printed by Printforce, the Netherlands